MICROELECTRONICS: A STANDARD MANUAL AND GUIDE

Other Books by
John Douglas-Young

Complete Guide to Electronic Test Equipment and Troubleshooting Techniques

Complete Guide to Reading Schematic Diagrams, Second Edition

Illustrated Encyclopedic Dictionary of Electronics

Practical Oscilloscope Handbook

Technician's Guide to Microelectronics

Microelectronics: A Standard Manual and Guide

JOHN DOUGLAS-YOUNG

Prentice-Hall, Inc.
Englewood Cliffs, New Jersey

Business and Professional Division

Prentice-Hall International, Inc., *London*
Prentice-Hall of Australia, Pty. Ltd., *Sydney*
Prentice-Hall Canada, Inc., *Toronto*
Prentice-Hall of India Private Ltd., *New Delhi*
Prentice-Hall of Japan, Inc., *Tokyo*
Prentice-Hall of Southeast Asia Pte. Ltd., *Singapore*
Whitehall Books, Ltd., Wellington, *New Zealand*
Editora Prentice-Hall do Brasil Ltda., *Rio de Janeiro*

©1984 by

PRENTICE-HALL, INC.

Englewood Cliffs, N.J.

Second Printing . . . . . June 1984

Editor: George E. Parker

Library of Congress Cataloging in Publication Data

Douglas-Young, John.
 Microelectronics: a standard manual and guide.

 Includes index.
 1. Microelectronics. I. Title.
 TK7874.D58   1984   621.381'7   83-9450

ISBN 0-13-581108-2

Printed in the United States of America

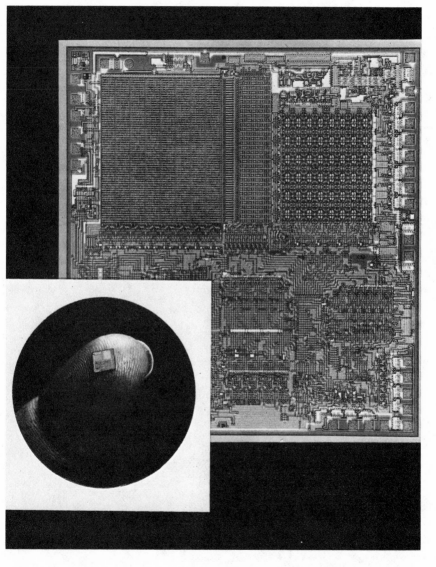

Computer-on-Chip

(Courtesy Rockwell International Inc.)

About This Book

Recently there has been a tremendous expansion in the design and fabrication of integrated circuits, and it appears as if manufacturers are now incorporating these devices into just about every type of electronics product being marketed today.

But the expansion so far will seem like a drop in the bucket compared with what is coming. For the last 20 years, all electronics products used in the exchange of information have been decreasing in price by a factor of 10 every five years despite inflation, and this trend is expected to continue for at least another ten years. Consequently, more and more features are being added to TV sets, radios, hi-fis, telephones, and so on, to increase their efficiency, reliability and attractiveness to customers; and these require the provision of an ever-widening variety of integrated circuits. This, in turn, makes possible the improvement or introduction of even more applications, such as appliances that talk to you and obey your verbal commands.

One of the most exciting directions that this trend is taking is toward digitization of home video and audio equipment. For instance, over 20 companies are marketing players for the new digitized audio disk which plays 15 times longer than the same size phonograph disk, and is completely noise-free. In TV, additional functions already available or coming shortly include microprocessor-controlled tuning, time-of-day display, devices for interfacing with various information networks such as Teletext and Videodata, and many more features, as manufacturers vie with each other for the business. In a year or two from now, color sets will require only one or two chips to provide all circuits except the power supply and the high voltage (the picture tube will be around for some time yet). Video recording will become more and more popular as prices come down, and a miniaturized system built into a video camera may eventually replace the home movie outfit. Microprocessor-controlled car radios will also be available, along with front radar speed control, rear sonar backup warning, electronic navigation, and so on. By the late 1980s the car will *almost* drive itself. One day it will!

7

What does all this mean to you and to others like you, who are going to have to cope with all this? Will you be ready to take full advantage of the opportunities offered by the almost limitless future?

This question is addressed in particular to those who were reared on analog theory and discrete circuitry, and have not as yet switched to digital thinking. The shock waves of these newer and more advanced developments are beginning to rattle our profession again just as it was settling down after the previous earthquake! Yet any technician who does not update his or her knowledge constantly will become obsolete very shortly. In this highly volatile technology giant forward leaps can come at any time, and wait for no one!

In *Microelectronics: A Standard Manual and Guide*, we included everything to bring you up to date and give you a thorough grounding in the fundamentals of digital theory. At the same time, we kept it simple and avoided difficult mathematics, for you are a busy person who wants *practical* knowledge above all. The two main technologies of microelectronics—unipolar and bipolar—are explained in detail, and so are its two major applications—digital and analog. Right . . . we didn't abandon linear circuits by any means, in spite of the tide that is setting in toward digitization. In fact, we included in the linear part of the book an explanation of the digitization of analog signals and vice versa.

The rapidly developing optoelectronics technology is important enough to be covered in a chapter of its own. Optical links are beginning to replace coaxial cable because of the larger bandwidth of optical fibers. These, in turn, require new optical sources and detectors, as well as new modulation and demodulation techniques. They open up an exciting future for cable services of all kinds. These and new types of display, even the solid-state picture screens that are bound to come eventually, are described and explained.

Another chapter is set aside for the discussion of printed circuit boards. These require some care in the mounting and demounting of components, especially ICs. You are told of the precautions you must take in handling, fabrication and repair to obtain successful results.

Logical procedures for troubleshooting are a must, and these are set forth in clear, easy-to-understand language for both digital and linear equipment. To those who have not had to troubleshoot digital equipment before, the chapter on this subject will be a revelation, and worth the price of the entire book.

There are some extremely useful Appendices, which include a glossary of microelectronic terms and abbreviations, a conversion table for metric and conventional units, all the electronics formulas and mathematical tables you probably will ever need, the standard graphic symbols for semiconductor devices, and a straightforward explanation of the binary number system.

This up-to-the-minute manual and guide will prove to be extremely valuable in meeting your need for quick, effective upgrading in this fast-moving technology. Although the current never stops, we believe that this book, coinciding as it does with today's and tomorrow's explosive application of large-scale integration, digitization and optoelectronics, will be comprehensive enough to carry you through the rapids, and bring you greater security and success in your professional future!

John Douglas-Young

TABLE OF CONTENTS

About This Book   7

Chapter 1   Microelectronics, Alias Integrated Circuits   21

Microelectronics and Integrated Circuits • Semiconductors • Conduction in Semiconductors • Impurities • Field-Effect Transistor (FET) • Basic Structure of Enhancement-Mode MOSFET • How the Field Effect Is Created • Depletion-Mode MOSFET • Junction Field-Effect Transistor (JFET) • Carrier Mobility • Current-Voltage Characteristics of MOSFETs • Transconductance of MOSFETs • Frequency Response of MOSFETs • Bipolar Junction Transistor • Junctions • How a BJT Amplifies • Common-Emitter Circuit • Frequency Response of BJTs • Basic Structure of a BJT • Resistors in ICs • Capacitors in ICs • Inductors in ICs • Fabrication of Integrated Circuits • Preparation of Silicon Wafer • MOS Fabrication • Bipolar Fabrication • Ion Implantation • Electron-Beam Machining • Performance Testing • Separation, Mounting, Bonding, Packaging

Chapter 2   Family Background Is Important   57

Digital Integrated Circuits • NOT, AND, and OR Gates • NAND and NOR Gates • Exclusive OR and NOR Gates • Threshold Voltage • Noise Margin • Propagation Delay • Fan-In and Fan-Out • Logic Families • DCTL • RTL • DTL • TTL • CML(ECL) • I^2L • STL • PMOS • NMOS • CMOS • SOS • MOSFET Versus BJT

Chapter 3   The Data Goes Round and Around **85**

Bistable   Multivibrator • R-S   Flip-Flop • D   Flip-Flop • J-K
Flip-Flop • J-K Master-Slave Flip-Flop • Clock • Shift Register
• Counter • BCD Counter • Buffer • Encoders • Decoders
and Display Drivers • Multiplexer • Timer

Chapter 4   Chips for the Memory **105**

Static RAM • Dynamic RAM • ROM, PROM • EAROM •
Charge-Coupled Device (CCD) • Bubble Memory

Chapter 5   The Microprocessor Directs the Traffic **113**

The Microprocessor • A Computer's IQ Is Zero • The Central
Processing Unit • I/O Devices • Diskette Controller • Keyboard Interface • Video Display Interface • Printer Interface

Chapter 6   Most Amplifier ICs Are Differential or
                Operational **129**

Linear Amplifiers • Direct-Coupled Amplifiers, Including Darlington-Coupled Amplifiers • DC Drift • Differential Amplifiers • Common-Mode   Rejection • Operational   Amplifiers •
Wideband Video Amplifiers • Narrowband IF Amplifiers •
Differential Comparators • Industrial Controls • Power Amplifiers

Chapter 7   But Today's Linear ICs Include More Than
                Amplifiers **145**

Video and Sound Detectors • Phase-Locked Loop (PLL) •
Voltage-Controlled   Oscillator   (VCO) • Rectifiers • Voltage
Regulators • Analog-to-Digital   Converter • Digital-to-Analog
Converter • Active Filter • Feedback Volume-Control Circuit
• Integrated Circuit Diode Array •

Chapter 8   Microwaves of the Future **159**

Conventional Radar System • Microelectronic Radar System •

Transmission   Lines • Varactors • IMPATTs • Gunn-Effect
Diodes • LSA Diodes • Phased-Array Antennas • Pulse-Com-
pression Filters • Circulators • Delay Lines • Phase Shifters •
Consumer Microwaves

Chapter 9   The Optoelectronics Connection **175**

Optical Communications • Optical Spectrum • Fiber Optics •
Lasers • Light-Emitting Diodes • PIN Diodes • Modulation
and Demodulation • Integrated Services Digital Network •
Displaying Information • LEDs • Liquid-Crystal Display
(LCD) • Vacuum-Fluorescent Display • Gas-Discharge Display
• Cathode-Ray Tube • Optoisolators

Chapter 10   Making Printed Circuit Boards Is Easy and Fun . **195**

Types of Boards • Factors Affecting Choice • Manufacturing
Procedures for PCBs • Making A PCB • Perfboards • Repair-
ing a Printed Circuit Board

Chapter 11   Troubleshooting Digital IC Equipment **207**

Why Troubleshooting ICs Is Different • Common Defects •
Using a Logic Probe • Using a Logic Pulser • Using a Logic
Clip • Using a Logic Comparator • Determining Circuit in
Trouble • Special Problems in Troubleshooting Microprocessor
Equipment • Logic State Analyzers

Chapter 12   Troubleshooting Linear IC Equipment **223**

General Troubleshooting Technique • Visual Inspection • Con-
firm Complaint • Resistance Checks • DC Voltage Analysis •
Signal Tracing • Waveform Analysis • Signal Injection • Brute
Force • Part Substitution • Component Tests • Alignment •
Testing Operational Amplifiers

Appendix 1   Glossary of Microelectronic Terms **235**

Appendix 2   Conversion Factors (U.S. and Metric) **249**

Appendix 3   Electronics Formulas and Mathematical Tables . 251

Appendix 4   Standard Semiconductor Symbols 271

Appendix 5   Binary Number System 273

Index 279

LIST OF ILLUSTRATIONS

Figure	Title	Page
Frontis-piece	Computer-on-Chip	5
1.1	(a) P-channel enhancement-type MOSFET, with single gate, active substrate, symbol; (b) P-channel enhancement-type MOSFET, structure	26
1.2	Depletion-mode MOSFET	28
1.3	Junction field-effect transistor	29
1.4	Miller indices	31
1.5	MOSFET current/voltage characteristics	33
1.6	NPN junction transistor	36
1.7	Point-contact transistor	37
1.8	Basic transistor circuits	38
1.9	Family of characteristic curves for BJT in common-emitter circuit	38
1.10	Common-emitter circuit for BJT in Figure 1.9	38
1.11	(a) Basic structure of BJT; (b) Making a diode out of transistor in Figure 1.11(a)	41
1.12	Czochralski method of crystal pulling	43
1.13	Cylinder of single-crystal silicon	44
1.14	Window etching	45
1.15	Bipolar fabrication	48
1.16	Packaging	51
1.17	Thermocompression bonding	55
2.1	Three basic logic gates: (a) AND; (b) OR; (c) NOT	58
2.2	NOT gate, or inverter	59
2.3	CMOS inverter	60
2.4	Gate protection circuit	61
2.5	AND gate	62
2.6	OR gate	63
2.7	(a) NOR and NAND gates; (b) Exclusive OR and NOR gates	64

Figure	Title	Page
2.8	Inverters cascaded	65
2.9	Direct-coupled circuit characteristics	66
2.10	Basic DCTL NOR gate	68
2.11	RTL NOR gate	69
2.12	DTL NAND gate	71
2.13	TTL NAND gate	72
2.14	CML (ECL) NOR gate	73
2.15	I²L NOR gate, circuit	74
2.16	I²L NOR gate, construction	75
2.17	STL	76
2.18	CMOS	80
2.19	SOS	81
3.1	Bistable multivibrator	86
3.2	R-S flip-flop with inverters	86
3.3	R-S flip-flop with two NAND gates	86
3.4	R-S flip-flop with four NAND gates	87
3.5	D flip-flop	87
3.6	D flip-flop timing diagram	88
3.7	D flip-flop with set and reset inputs	88
3.8	J-K flip-flop	88
3.9	J-K master-slave flip-flop	89
3.10	Four-bit serial-in/parallel-out shift register	91
3.11	(a) Counter circuit; (b) Timing diagram	93
3.12	(a) BCD counter; (b) Timing diagram	95
3.13	Five BCD counters give a five-figure readout	96
3.14	Buffer	97
3.15	Encoder	98
3.16	Decoder	99
3.17	Multiplexer	100
3.18	Timer (one-shot)	101
3.19	Timer (free-running)	102
4.1	Static RAM	106
4.2	Dynamic RAM	107
4.3	MOS ROM matrix	108
4.4	Charge-coupled device	110
4.5	(a) Portion of bubble-memory chip; (b) Permalloy tracks	111
4.6	Principle of bubble memory	112
5.1	Basic functions of a computer	114
5.2	Central processing unit	115

Figure   **Title**   **Page**

5.3   Basic microcomputer diagram   116
5.4   Timing diagram for microcomputer in Figure 5.3   117
5.5   Computer on a chip   118
5.6   (a) Sector format; (b) Diskette controller IC   121
5.7   (a) Keyboard operation; (b) Interface IC   123
5.8   (a) Displayed character; (b) CRT controller interface   124
5.9   (a) Dot matrix character; (b) Printer interface IC   127

6.1   Linear amplification   130
6.2   Two-stage discrete transistor amplifier   130
6.3   Direct-coupled transistor amplifiers   131
6.4   Differential amplifier stages   132
6.5   General purpose operational amplifier   134
6.6   Operational amplifier circuits   135
6.7   Operational amplifier computing functions   136
6.8   Various op-amp circuits   138
6.9   Circuit diagram for wideband video amplifier   139
6.10   Typical industrial control centers 142

7.1   Video detector circuit 146
7.2   Multimode detector IC 146
7.3   Phase-locked loop 147
7.4   Partial schematic of VCO 148
7.5   Basic voltage regulator IC   149
7.6   A-D converter for DMM 150
7.7   (a) Dual-ramp A-D converter; (b) Dual-ramp waveforms . .   151
7.8   Conversion of audio to PCM 152
7.9   D-A converter 153
7.10   Active filter IC 153
7.11   Feedback volume control 154
7.12   IC diode array 155
7.13   Diode array used in limiting circuit   155
7.14   Diode array used in balanced modulator circuit   156
7.15   Diode array used in balanced mixer circuit   156
7.16   Diode array used in ring modulator circuit   157

8.1   Conventional radar system   160
8.2   Microstrip structure and characteristic impedance   161
8.3   Microacoustic line 162
8.4   IMPATT or varactor diode mounted on stripline   164
8.5   Gunn-effect diode 165
8.6   Gunn-effect current waveform   166

Figure	Title	Page
8.7	Principle of solid-state phased array radar antenna	167
8.8	Microelectronic radar system	169
8.9	Circulator	170
8.10	Microacoustic interdigital structures with nonuniform grating can be used as dispersive delay lines	170
8.11	Phase shifter	171
9.1	Propagation in single-mode fiber	178
9.2	Propagation in multimode fiber	178
9.3	Pulse broadening in multimode fiber	178
9.4	Graded-refractive-index fiber propagation	178
9.5	Delta modulation	179
9.6	Maximum acceptance angle	179
9.7	Injection laser	181
9.8	Light-emitting diode with microlens	182
9.9	PIN diode	183
9.10	Modulation circuit for laser	184
9.11	Demodulation circuit for PIN diode	185
9.12	Integrated Services Digital Network	186
9.13	Solid-state displays	187
9.14	Square-wave phasing for LCD	190
9.15	Solid-state display screen	192
9.16	Optoisolator	194
10.1	Printed-circuit board	197
10.2	Mounting components on a perfboard	204
11.1	Effect of open bond on NAND gate input	209
11.2	Logic probe	211
11.3	Logic clip	213
11.4	Circuit diagram of logic chip	214
11.5	Oscilloscope with logic analyzer	219
11.6	Two-state logic analyzer	220
12.1	Signal tracing technique	228
12.2	Testing an operational amplifier	232
IV.1	Graphic symbols	271
V.1	Half adder	276
V.2	Four-bit parallel adder	276

MICROELECTRONICS:
A STANDARD MANUAL
AND GUIDE

CHAPTER 1

Microelectronics, Alias Integrated Circuits

Microelectronics and Integrated Circuits

Microelectronics is the science dealing with the theory, design, and application of **integrated circuits.** Integrated circuits (ICs) are electronic devices containing many interconnected circuit elements formed on a single body, or **chip** of **semiconductor** material (see Frontispiece). Since the chip is the foundation of the device, it is also called the **substrate.**

Semiconductors

Since semiconductors are basic to ICs, we'll start by briefly reviewing some facts about them. In doing this we'll assume you have a general idea of the modern theory of atomic structure.

Silicon and **germanium** are semiconductors. They are chemical elements whose atoms have four **electrons** in their outer **shells.** These outer electrons are called **valence electrons,** and they provide the "glue" that holds the atoms together in a manner called **covalent bonding.** Covalent bonding occurs when valence electrons are shared between nearest neighbors. The bonds consist of two electrons, one from each neighboring atom. Since eight electrons are required for a closed outer electron shell, the arrangement of the atoms is such that each has four nearest neighbors. This sharing of electrons decreases the energy of the associated atoms as compared with the energy of

21

atoms when they are free, and this decrease provides the cohesive energy for the system, which may be conceived of as a crystalline solid made up of a vast number of infinitely small cubes packed together, like a Rubik's cube, in a **crystal lattice.**

Since the valence shells of the atoms in the crystal are interacting, these shells are obliged to split into many discrete energy levels, because no two electrons can be at exactly the same energy level in the same system. As each atom's valence shell contains four electrons, the total number of energy levels in the **valence band** is very great. Each atom also contains four energy levels higher than the valence band, but no electrons to occupy them. The total number of these energy levels makes up the **conduction band.** Between these bands is a region of energy in which electrons are not allowed. This is called the **forbidden band** or **energy gap.** At room temperature the gap is 1.08 electron volts "wide" in silicon; in germanium it is 0.65 eV.

There are also some semiconductors consisting of compounds of atoms with three and five electrons each. An example is **gallium arsenide,** in which each gallium atom (three valence electrons) is surrounded by four neighboring atoms of arsenic (five valence electrons each). The manner in which this semiconductor operates is similar in principle to silicon and germanium.

Conduction in Semiconductors

Semiconductors are classified as **intrinsic** or **extrinsic.** An intrinsic semiconductor is a crystal in which conduction occurs because a number of valence electrons have acquired enough energy to enable them to jump to the conduction band, leaving a corresponding number of broken valence bonds in the valence band. An extrinsic semiconductor is a crystal in which most of the conduction occurs because of **impurities** inserted into the crystal when it was formed.

At **absolute zero** all valence electrons are in their valence-band holes, since none have the energy to leave them. If a weak **electrostatic field** is applied across the semiconductor, no current will flow —at absolute zero the semiconductor is an insulator. At higher temperatures the atoms acquire **thermal energy** from their surroundings, and some electrons may reach the level necessary to hop out of their holes and into the conduction band. If an external field is applied now, the free conduction electrons will be accelerated toward the positive electrode of the device creating the field.

Where a hole was left in the valence band, it is possible, under the influence of an external field (or random thermal energy), for a valence electron from another atom to jump into the hole of the first atom. This hole disappears, but a new one appears in the valence band of the second atom. A valence electron from a third atom could similarly fill the hole in the second atom, leaving an empty hole in the third atom, and so on. Thus, there are *two* currents of electrons flowing toward the positive electrode: fast-moving free electrons flying through the spaces in the crystal lattice (but with numerous collisions), and slower-moving valence electrons hopping from hole to hole. However, the second current is conventionally conceived of as a current of **holes** flowing in the opposite direction.

How can this be? Well, it can be visualized easily if you arrange a few coins in a row so that they touch each other. Then move the left-hand coin an inch to the left. There is now a gap between the first coin and the second. Move the second coin up to the first and the gap "moves" to a position between the second and third coins. Continue moving the coins until the gap has traveled all the way to the right. Note: Each coin moved one space to the left, but the gap moved along the entire row. In the same way, each valence electron moves one step and stops, but the hole appears to move continuously in the opposite direction.

Impurities

The type of conduction described so far is carried on by the charge carriers (electrons and holes) generated in pure (intrinsic) silicon and germanium. However, practical devices require some modification of the characteristics of these semiconductors, as we shall see later. This modification is done by adding small amounts (about one part in a million) of other elements that create additional carriers. "Doped" (extrinsic) semiconductors are called **n type** if the additional carriers are negative (electrons), or **p type** if the additional carriers are positive (holes).

Impurities that may be used to obtain n-type material are **antimony, arsenic,** and **phosphorus.** Their atoms have *five* valence electrons, so when one of them substitutes for a germanium or silicon atom in the crystal lattice, four of its valence electrons form covalent bonds with four neighboring atoms (as in intrinsic material), but the fifth is free to become a conduction electron. Consequently, there is a substantial increase in conduction electrons, which are therefore the **majority carriers**. However, hole current also exists, but as the number of holes is no longer equal to the number of electrons, the holes are called **minority carriers**.

Impurities that may be used to obtain p-type material are **aluminum, boron,** and **indium**. Their atoms have *three* valence electrons, so when one of them substitutes for a germanium or silicon atom in the crystal lattice, all three valence electrons form covalent bonds, but a hole is created because the atom did not have a fourth valence electron. Consequently, there is a substantial increase in holes, which are therefore the majority carriers while electrons are the minority carriers.

Impurities that contribute additional free electrons are called **donor** impurities. Conversely, those that contribute holes are called **acceptor** impurities. You should remember, however, that the additional carriers do not alter the overall charge on the material, since a donor atom's nucleus has a positive charge equal to the negative charge removed by the freed electron, and an acceptor atom that fills its extra hole with an electron from a nearby silicon or germanium atom acquires thereby a negative charge.

Another point to bear in mind is that the number of carriers increases with temperature. This means that current increases with temperature. But then, so does the number of collisions made by free electrons with the crystal lattice. The absorbed energy from these collisions raises the lattice temperature further, thus freeing even more carriers. This is a runaway situation that would lead to destruction of the semiconductor material if not controlled by appropriate protective circuitry.

Semiconductor devices used in ICs may make use of one type of carrier or both, and are called **unipolar** or **bipolar** accordingly. We shall begin with the unipolar type.

Field-Effect Transistor (FET)

The field-effect transistor was patented in the United Kingdom by Oscar Heil in 1935, but remained dormant until the development of ICs in the 1960s. At this time, the need for even smaller transistors and simpler manufacturing processes, together with the availability of commercial quantities of silicon of sufficient purity, brought about its resurrection.

In principle, the FET is like a vacuum tube in that it is a voltage-operated device with very high input impedance. The flow of direct current is controlled by an electric field emanating from a **gate** electrode. This field is created by the voltage on the gate, and varies in intensity with that voltage. There are two main types of FET:

Insulated-gate FET (IGFET)

Junction FET (JFET)

The IGFET, in turn, is divided into two main types:

Enhancement mode

Depletion mode

Any of these may be fabricated on either p-type or n-type material. The enhancement-mode device is the one preferred for ICs because in the absence of a gate signal it draws no current. The depletion-mode device is just the opposite. Because of its structure, an IGFET is commonly called a **metal-oxide-semiconductor field-effect transistor (MOSFET or MOS/FET)**.

Basic Structure of an Enhancement-Mode MOSFET

Figure 1.1 shows the standard symbol* (a) and basic structure (b) of an enhancement-mode p-channel MOSFET. In this MOSFET there are two p-type regions formed by diffusing boron into an n-type silicon substrate. The two regions are called the source and the drain. The n-type space (L) between them is termed the channel and the dimension L is typically 25 μm.† A very thin dielectric layer of silicon dioxide (SiO_2), nominally 0.1 μm thick, insulates the channel from an aluminum electrode called the gate.

---

*Other MOSFET symbols are given in the Appendix.

†Metric units are used throughout this book. A conversion table is given in the Appendix.

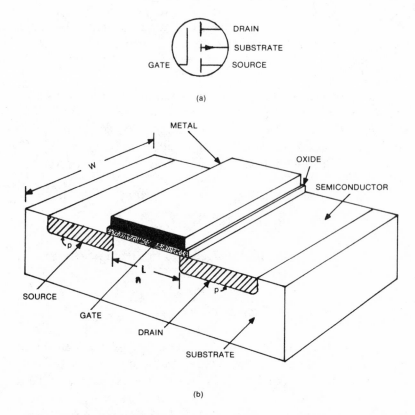

Figure 1.1 (a) P-channel enhancement-type MOSFET, with single gate, active substrate (envelope symbol may be omitted if no confusion will be caused); (b) P-channel enhancement-type MOSFET structure.

Since boundaries between p and n regions are junctions, the two p regions and the n region between them comprise two diodes, back to back. When a battery is connected between source and drain, one of these diodes will be reverse biased whichever way the battery is connected, so no current will flow. (Diodes are discussed below under bipolar devices.)

How the Field Effect Is Created

In a semiconductor at room temperature hole-electron pairs are continually being generated and recombined. Electrons are the majority carriers in n-type silicon, but if a sufficiently negative voltage is applied to the gate electrode, the resulting field repels the negative majority carriers from the vicinity of the silicon-silicon dioxide interface, causing a **depletion layer** populated only by positive minority carriers, hence the term **p-channel**. In effect, the n-type semiconductor has changed to p type, so it is also called the **inversion layer**. The p-n junctions now no longer exist where the inversion layer meets the source and drain, so conduction by holes may take place.

The inversion layer is extremely thin ($\approx$ 5 nm). Its depth varies with the electric field strength, which in turn depends on the gate potential (V_G). The electric field does not penetrate much into the gate electrode because the free carrier density of metal is very high ($10^{22}/cm^3$), but the free carrier density of the semiconductor is considerably lower ($10^{16}/cm^3$). Consequently, the lines of force of the field have to penetrate the silicon to a greater depth in order to terminate on positive charges, so the more of them there are, the further some of them will have to reach.

A certain minimum gate potential is required to get things started. This is the voltage necessary to cancel various small built-in charges at the interface. As we shall see later on, these depend on the physical characteristics of the gate. This potential is called the **threshold voltage** (V_T). An average value for the basic enhancement-mode MOSFET is -4 V. When $V_G = V_T$ the MOSFET is ready to turn on. Without a gate potential, no conduction can take place, so it is normally *off*. Conduction increases with field intensification, hence the term enhancement-mode.

Depletion-Mode MOSFET

Figure 1.2 shows a **depletion-mode** MOSFET. In this device the substrate is p-type silicon. Phosphorus is diffused into the p-type surface to form an n-type region. This region comprises the source, drain and n-channel between, so there are no p-n junctions. When a battery is connected between source and drain, current flows continuously. This device, therefore, is normally *on*.

A negative potential applied to the gate electrode has the same effect on the n-type region beneath it as in the enhancement-mode MOSFET. Negative carriers are repelled and positive carriers attracted, until an inversion layer is created. The deeper this goes, the more the n region becomes a p region, and the n channel be-

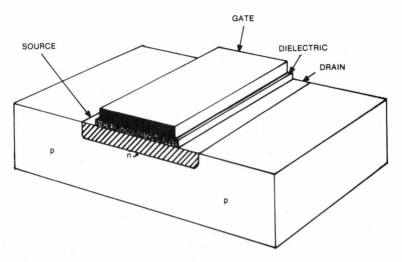

Figure 1.2   Depletion-mode MOSFET.

comes increasingly depleted of negative carriers. Eventually, conduction through the channel is entirely cut off. The gate potential required to bring this about is called the **pinch-off voltage** (V_p).

Both types of MOSFET have bilateral symmetry, therefore conduction can take place in either direction. Current is assumed to come from the source and leave by the drain, so if conduction is by holes, the drain will be more negative than the source, and vice versa if conduction is by electrons.

Junction Field-Effect Transistor (JFET)

Figure 1.3 shows the construction of an n-channel JFET. Its channel consists of a very thin plate of n-type semiconductor, with gate pads formed on the upper and lower surfaces by doping the areas to make them p-type. The **junctions** thus formed between the p-type and n-type material are reverse-biased by the negative voltage from battery V_{gg}, so no current flows from the gate regions to the channel. (Junctions are discussed in more detail below under **bipolar transistors**.)

The negative potential on the gate electrodes repels the nega-

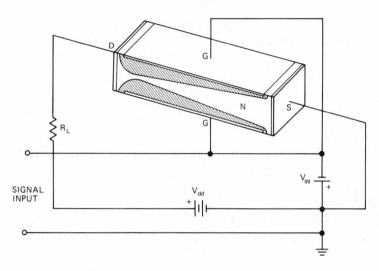

Figure 1.3 Junction Field-Effect Transistor.
D = drain (ohmic contact)
G,G = gate (p-type)
N = channel (n-type)
S = source (ohmic contact)
Shaded area = depletion region. The depletion region becomes larger as V_{gg}
increases. If V_{gg} increases enough the channel is "pinched off."

tive charge carriers in the n-type channel, causing depletion regions (shaded areas). The depth of these regions varies with the gate voltage, so that the current flowing from source to drain is reduced as the gate voltage goes more negative, and vice versa. If the gate voltage goes negative enough, the channel is blocked entirely and no current flows. The voltage at which this happens is called the pinch-off voltage (V_p), as in the depletion-mode MOSFET.

A JFET can also be constructed with a p-type channel and n-type gates. Used with supply voltages of opposite polarity to those required by the n-type channel JFET, its operation is just the same, except that holes form the channel current.

Carrier Mobility

An electron in free space when subjected to an electric field undergoes continuous acceleration, but when in a semiconductor, it can only reach a limited velocity because of collisions with imperfections in the crystal lattice and encounters with ionized impurities. The same thing applies to holes. The actual velocity attained is termed the **drift velocity** (v), and depends on the strength of the electric field (ε) and the limiting velocity imposed by the semiconductor. The latter varies according to the density of lattice imperfections and concentration of dopant, and is called the **carrier mobility** (μ). Therefore:

$$v = \mu \varepsilon \qquad\qquad (1)$$

If the velocity is in centimeters per second and the electric field is in volt per centimeter,

$$v = cm/s \qquad\qquad (2)$$

$$\varepsilon = V/cm \qquad\qquad (3)$$

Rearranging (1) to get μ gives

$$\mu = v/\varepsilon \qquad\qquad (4)$$

and substituting (2) and (3) for v and ε gives

$$\mu = cm/s \div V/cm$$
$$= cm/s \cdot cm/V$$
$$= cm^2/V \cdot s \qquad\qquad (5)$$

which is the unit of carrier mobility.

As we noted above, under semiconductors, the way in which the atoms are arranged in single-crystal silicon results in what is most easily visualized as a three-dimensional lattice of small cubes all oriented the same way. When a slice of this structure is severed from the parent block, the cut surface will consist of a mosaic of truncated cubes. Since the charge carriers in the channel of a MOSFET are obliged to travel in the inversion layer, which is only about five nanometers thick, their mobility will vary with the texture of the surface layer, the characteristics of which will depend on the angle the surface plane makes with the cubes it cuts.

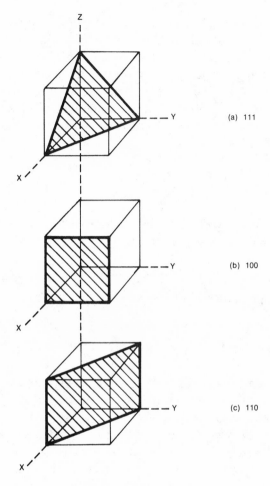

Figure 1.4 Miller Indices.

This is illustrated in Figure 1.4, where single cubes are shown cut in different ways. Each cube has three axes labeled X, Y and Z. In (a) all three axes have been cut. With this angle of cut the silicon surface will consist of a layer of pyramidal segments. In (b) only one axis has been cut, so this surface will consist of rectangular blocks. In (c) two axes have been cut, which will give a surface of triangular prisms.

In order to specify the plane of the surface, a **Miller index** is used. This is a three-digit number in which each axis cut is denoted by a figure 1. When the plane cuts three axes as in (a) the index is (111). In (b), where only one axis is cut, the index is (100). In (c) the index is (110), which tells us that two axes are cut.

As a consequence, carrier mobility in the standard p-channel enhancement-mode MOSFET is only about half of what it is for holes in a junction transistor with the same type of silicon. The actual values for the (111) and (100) planes (which are those generally used) at 27° C are:

$$\mu_p (111) \approx 190 \text{ cm}^2/\text{V} \cdot \text{s}$$
$$\mu_p (100) \approx 130 \text{ cm}^2/\text{V} \cdot \text{s}$$

In n-channel devices, where electrons are the carriers, the greatest mobility is in the (100) plane:

$$\mu_n (100) \approx 600 \text{ cm}^2/\text{V} \cdot \text{s}$$

which is approximately three times the hole mobility in the (111) plane.

Current-Voltage Characteristics of MOSFETs

A graph of drain voltage (V_D) versus drain current (I_D) for various levels of gate voltage (V_G) is shown in Figure 1.5. The dashed curve indicates where $|V_G - V_T| = V_D$. This is the boundary between the linear (or "triode") region to the left and the saturation region to the right. In the linear region $|V_G - V_T|$ is greater than V_D, and the current-voltage characteristics resemble those of a vacuum-tube triode. In the saturation region, $|V_G - V_T|$ is less than V_D. Current cannot be increased by increasing V_D because the channel is saturated. Only by deepening the channel (by increasing V_G) can I_D be made any greater.

Transconductance of MOSFETs

Transconductance (g_m) in the "triode" region is given by:

$$g_m = \mu \frac{C_g}{L^2} |V_D| \tag{6}$$

where: μ = carrier mobility (cm^2/V · s)

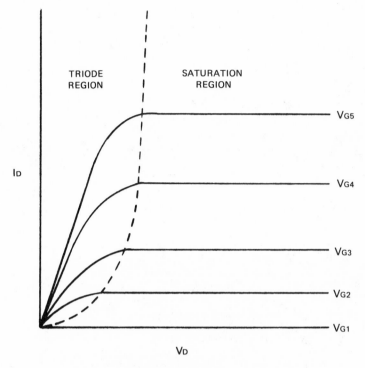

Figure 1.5 MOSFET current/voltage characteristics.

C_g = total gate-to-channel capacitance in farads

L = channel length in centimeters

V_D = drain-source voltage in volts

For example, if μ = 190 cm^2/V · s [for the (111) plane]

$$C_G = 9 \text{ pF}$$

$$L = 0.001 \text{ cm } (10 \ \mu\text{m})$$

$$V_D = 1 \text{ volt}$$

$$g_m = \frac{190 \times 9 \times 10^{-9}}{(0.001)^2} \times 1$$

$$= 1.71 \text{ millisiemens (or mS)}^*$$

*Millisiemens is the metric replacement for millimhos.

In the saturation region transconductance is given by:

$$g_m = \mu \frac{C_g}{L^2} |V_G - V_T| \tag{7}$$

Frequency Response of MOSFETs

The maximum frequency of operation (f_o) of a MOSFET is given by:

$$f_o = \frac{g_m}{2\pi C_g} \tag{8}$$

Combining equations (7) and (8):

$$f_o = \frac{\mu |V_G - V_T|}{2\pi L^2} \tag{9}$$

For example, where $V_G = -10$ V

$$V_T = -4 \text{ V}$$
$$L = 0.0005 \text{ cm } (5 \ \mu m)$$
$$\mu = 190 \text{ cm}^2/\text{V} \cdot \text{s}$$
$$f_o = \frac{190 \times 10^{-4} \times 6}{2\pi \times (5 \times 10^{-6})^2}$$

$$\approx 726 \text{ MHz}$$

Unfortunately this high cut-off frequency is theoretical only. Distributed capacitance on the chip reduces it by a factor of ≈ 100.

Bipolar Junction Transistor

So far we have been considering unipolar devices, but it is now time to turn our attention to the competition. The **bipolar junction** transistor (BJT) uses both types of charge carrier simultaneously, and has two junctions.

Junctions

A junction is a boundary between n-type and p-type regions in a semiconductor crystal. Initially, the majority carriers in each region diffuse across the boundary, spreading out evenly throughout the crystal, and a large number neutralize each other by recombination. This reduces the number of free electrons and holes to the point where the p-type region becomes *negative* and the n-type region becomes *positive*.

If a battery is now connected across the crystal so that its positive terminal is connected to the n-type region and its negative terminal to the p-type region, the remaining majority carriers will be "sucked out," and the crystal becomes virtually an insulator, since there are no majority carriers left to carry the battery current. A small current of *minority carriers* exists, however.

On the other hand, if the connections are reversed, the battery floods the n-type region with electrons and the p-type region with holes. Current by majority carriers now flows freely across the junction.

This is the explanation of how a single junction is used in a **diode**. The n-type region is the **cathode** and the p-type region the **anode**. When the battery is connected with its negative terminal to the anode and its positive terminal to the cathode, the diode is **reversed biased**, and it is for all practical purposes an insulator. With the battery connections the other way, the diode is **forward biased** and becomes a conductor.

In the case of a BJT, there are two junction diodes in series but with opposite polarity. That is to say, they are arranged in a **pn-np** sequence of doped regions, or an **np-pn** sequence. However, instead of being separate diodes, they are combined in a **pnp** or **npn** sandwich. The included n or p region is also made very thin.

Figure 1.6 shows an npn BJT connected so that the left-hand "diode" is forward biased and the right-hand "diode" reverse biased. Majority carriers flow readily across the left-hand junction, but not across the one on the right. In the n region on the left electrons are

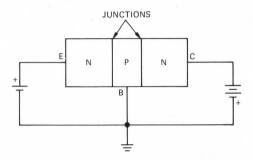

Figure 1.6 NPN Junction Transistor. B = base; C = collector; E = emitter; N = n-type; P = p-type semiconductor.

majority carriers, so they are "injected" by the forward bias into the p region. But in the p region electrons are minority carriers; and because this region is very thin they soon find themselves within the electrostatic field of the other, reverse-biased junction. As mentioned above, minority carriers do flow across reverse-biased junctions; the current is negligible in a diode because the minority carriers are few. But these former majority carriers are numerous, and most of them are swept through the right-hand junction by the positive potential on the right-hand n region.

In a pnp BJT the action is similar, but the majority carriers injected from the p region into the n region are holes, and the biasing voltages are reversed.

How a BJT Amplifies

For reasons that are now obvious, the region that provides the majority carriers is called the **emitter**, and the region that receives most of them is called the **collector**. A small percentage of carriers do not reach the second junction, however, as they encounter majority carriers in the **base** and disappear by recombination. (The term *base* comes from the point-contact transistor, shown in Figure 1.7, which was the form of the first bipolar transistor.) The collector current, therefore, equals the emitter current less the base current, or:

$$I_c = I_c - I_b \qquad (10)$$

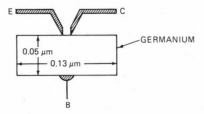

Figure 1.7 The earliest form of transistor was the point-contact transistor, invented at Bell Telephone Laboratories in 1948 by John Bardeen, Walter H. Brattain, and William B. Shockley. The term "base" (B in this sketch) arose from this electrode being the understructure of the device.

To illustrate how a BJT amplifies, take as an example one that has an I_c/I_e ratio (**alpha** or α) of 98. Since 98 percent of the emitter current goes into the collector, the base current must be 2 percent of the emitter current. Suppose I_b is initially 0.2 milliampere, and I_c is 10 mA. If we increase the emitter-base forward bias so that I_b increases to 0.3 mA, I_c must increase to 15 mA; and I_c, which was originally 9.8 mA, increases to 14.7 mA. The increase of 0.1 mA in the base current produced an increase of 4.9 mA in the collector current; or a current gain of $4.9/0.1 = 49$.

Common-Emitter Circuit

There are three basic transistor circuits, as shown in Figure 1.8. The one most commonly used in ICs is the **common-emitter circuit**. In this configuration, the input signal is applied between the base and the emitter and the output signal appears between the collector and emitter (hence the term, *common emitter*). The signal adds to and subtracts from the very small base current, and its variations cause proportionally larger variations in the collector current, as we have just seen.

In Figure 1.9, collector current (I_c) is plotted against collector voltage (E_c) for various values of base current (I_b) for a certain BJT. Suppose we want to use this BJT in the circuit given in Figure 1.10.

The first thing we notice is that the load resistor R_L has a value of 5000 ohms. The supply voltage V_{cc} is 20 volts, so the maximum collector current I_c is given by:

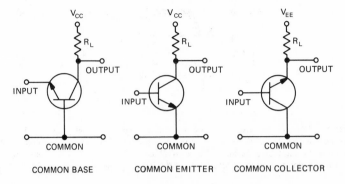

Figure 1.8 Basic Transistor Circuits.

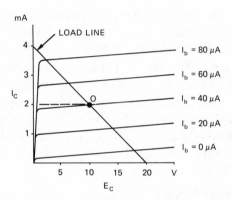

Figure 1.9 Family of characteristic curves for BJT in common-emitter circuit. O = operating point.

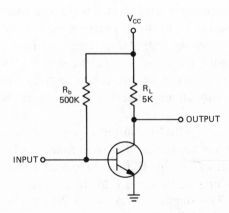

Figure 1.10 Common-emitter circuit for BJT in Figure 1.9.

$$I_c = 20/5000 = 0.004 \text{ A, or 4 mA}$$

On the graph in Figure 1.9, draw a line connecting this value (4 mA) on the vertical axis with the value of V_{cc} (20 volts) on the horizontal axis. This line is called the **load line**. A point midway along this line is called the **operating point**. In this case the operating point corresponds to an I_b of 40 μA; bias current can swing from zero to 80 μA on either side of this point for linear amplification.

At the operating point I_c is 2 mA and E_c is 10 volts. The power dissipated at the collector is therefore:

$$P_d = E_c I_c = 10 \times 2 \times 10^{-3} = 20 \text{ mW}$$

The bias resistor, R_b, is shown having a value of 500 kilohms. This value is obtained from:

$$R_b = V_{cc}/I_b = 20/40 \times 10^{-6} = 5 \times 10^5 = 500 \text{ k}\Omega$$

Figure 1.9 also tells us that if I_b increases to 80 μA, I_c must increase to 3.6 mA, and if I_b decreases to zero, I_c decreases to zero likewise. In other words, an input signal with a peak-to-peak current swing of 40 μA will give an output collector current swing of 1.6 mA peak-to-peak. The current gain G_c is therefore:

$$G_c = 1.6 \times 10^{-3}/40 \times 10^{-6} = 40$$

In the common-emitter circuit, the value of $\Delta I_c / \Delta I_b$* is called **beta** (β).

Frequency Response of BJTs

As with a MOSFET, the frequency response of a BJT is limited by its built-in capacitance. Each junction acts as a capacitor, and their combined effect shunts the input resistance. This can be ignored at low frequencies, but when the capacitive impedance equals the input resistance, the current-gain beta of the transistor is reduced

---

*Δ (delta) denotes increment or decrement of a value.

to 0.707 of the low-frequency gain. The frequency at which this oc-curs is called the **beta cut-off frequency** (f_β). You may have to calcu-late this from:

$$f_\beta = (1 - \alpha)f_\alpha$$

since manufacturers normally specify the **alpha cut-off frequency** (f_α).

Basic Structure of a BJT

Figure 1.11(a) illustrates the basic structure of a BJT in an IC. As you can see, it differs considerably from the MOSFET. This is an npn BJT, formed upon a p-type substrate. In the sectional draw-ing the vertical dimensions are grossly exaggerated. The n-type re-gions of the BJT are insulated from the p-type substrate because the junction between them is reverse biased.

The collector, emitter, and base are small n and p regions formed on the substrate by diffusing into it the appropriate impuri-ties, and they are connected to other circuit elements on the chip by a deposited metal interconnection pattern. Although it looks differ-ent from the diagram in Figure 1.6, it works in the same way. We'll be discussing the details of this type of fabrication later in this chap-ter.

Figure 1.11(b) shows how this transistor can be used as a diode.

Resistors in ICs

Resistance in ICs is provided by transistors. A conventional resistor fabricated from resistive material would be so large in com-parison with a transistor that it would take up too much room on the chip. This is particularly the case with MOS chips, where MOSFETs are even smaller than BJTs. For instance, an MOS load device of 100 kilohms occupies approximately one square mil. The same resistance value, using standard methods with sheet resistance typically in the range of 20 to 200 ohms per square, would obviously require a much larger area. A transistor, on the other hand, is like a variable resistor, and it can be adjusted to a fixed value by clamping

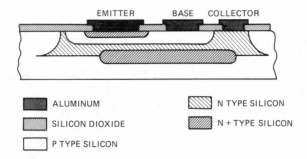

ALUMINUM

SILICON DIOXIDE

P TYPE SILICON

N TYPE SILICON

N + TYPE SILICON

Figure 1.11(a) Basic structure of BJT "island" (vertical section).

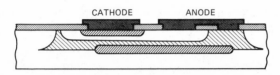

Figure 1.11(b) The transistor in Figure 1.11(a) can be made into a diode by shorting the base and collector together.

the base or gate with a set bias. For most purposes the tolerance (about 10 percent) of such "resistors" is adequate. For greater precision it is necessary to use discrete resistors which are connected externally. This is also necessary where resistance requirements are very high or very low.

Capacitors in ICs

In MOSFETs the gate is separated from the underlying semiconductor by a thin film of silicon dioxide. This makes a perfect capacitor. In bipolar devices a reverse-biased junction serves as a capacitor; the depletion region at the junction is the dielectric.

Actually, with MOS devices the problem is more often too much capacitance. A MOSFET contains three built-in capacitors. There is the gate electrode/substrate capacitor, with silicon dioxide as a dielectric; there is the drain/source capacitor with the channel as a dielectric; and there is the gate electrode/drain and source capacitor with silicon dioxide as a dielectric. There is, therefore, no problem about providing capacitance. As a rule the designer is more concerned with reducing it!

Inductors in ICs

Since it is manifestly impossible to build coils into an IC, the designer usually provides for inductance by the use of operational amplifiers with suitable feedback arrangements. External inductances are also frequently connected to the IC. In some cases inductance is obtained by using piezo or mechanical transducers.

Fabrication of Integrated Circuits

The first stage in the fabrication of MOS devices is the production of **single-crystal silicon**, in which the atoms are arranged in a perfectly regular pattern throughout. Silicon makes up 25 percent of the earth's crust, but seldom in a form that can be used in electronics. It is most often found combined with oxygen in the compound known as silica.

From silica we derive silicon tetrachloride, which is a liquid capable of being highly purified. When pure hydrogen is bubbled through this liquid it combines with the chlorine, reducing the silicon tetrachloride to pure **polycrystalline silicon**.

Preparation of Silicon Wafer

To convert it to single-crystal silicon, the polycrystalline silicon is melted in an inert gas in an airtight crucible heated to 1415° C by radio-frequency energy. At the same time donor or acceptor impurities are added, depending on which type of extrinsic silicon is desired.

When melted, the polycrystalline silicon loses its crystalline structure, but when a seed crystal of single-crystal silicon on the end of a rod is lowered on to it, the molten silicon is cooled below its melting point at the point of contact, and recrystallizes around the seed. The crystalline structure now is aligned with that of the seed. The rod is gradually raised, pulling up the crystallized silicon and allowing more to crystallize around it, so that little by little a cylinder of single-crystal silicon slowly emerges from the melt. This process, which is called the **Czochralski method**, is illustrated in Figure 1.12.

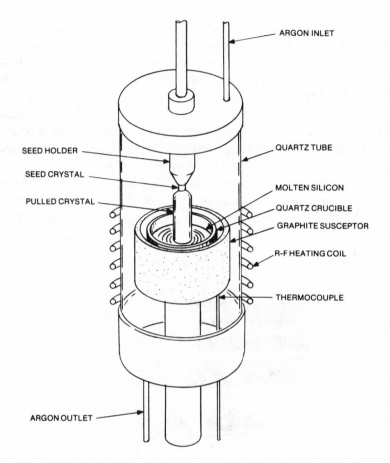

Figure 1.12 Czochralski method of crystal pulling. The rod rotates once per second and is slowly raised to form a silicon crystal about 5 cm. in diameter and 30 cm. long.

A second method of crystal pulling is known as the **floating zone method.** In this process a polycrystalline ingot of silicon is supported vertically, with a seed crystal of single-crystal silicon at its upper end. Localized RF induction heating is applied to heat the zone next to the seed crystal to its melting point. The molten zone is then made to travel to the lower end of the ingot as it is slowly raised. As each zone cools and recrystallizes, it does so as single-crystal silicon. By repeating the process several times, greater purity can be obtained. This method also avoids contamination that can arise from the use of a quartz crucible.

The cylinder of single-crystal silicon that results is some 50 millimeters in diameter. It is sliced with a diamond saw into disks of about one millimeter thick. The cut is made at the proper angle to give the required Miller index, and the slices are then subjected to various grinding, lapping, and polishing operations in which they acquire a mirror-like surface and are reduced to a thickness of some 200 micrometers (Figure 1.13). Each is then inspected under a low-power microscope for flatness, and also tested for crystal dislocation and electrical resistance in ohms per square centimeter.

Each of these slices will finally make a large number of individual chips, but they are not separated until the end of the process, since the wafers are much easier to handle than the tiny pieces of silicon. But considerable care is still needed, because silicon is extremely brittle.

Figure 1.13 Cylinder of single-crystal silicon is sliced into wafers with a diameter of 50 mm. and a thickness of 200 μm. after grinding, lapping and polishing.

MOS Fabrication

The first step (see Figure 1.14) is the formation of a layer of **silicon dioxide** on the surface of each wafer. Silicon dioxide (SiO_2) is very like ordinary glass, and makes a good insulating and protective surface covering. The batch of wafers is placed in an oven at a temperature of 1000° to 1300° C, and exposed to a current of dry oxygen. This combines with the silicon to form a layer of SiO_2 between 0.7 and 1.8 micrometers thick.

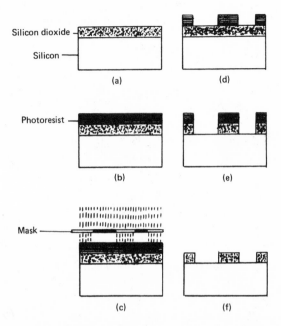

Silicon dioxide

Silicon

(a)

(d)

Photoresist

(b)

(e)

Mask

(c)

(f)

Figure 1.14 Window etching. (a) Wafer with SiO$_2$ layer. (b) Photoresist applied. (c) Photoresist exposed to ultraviolet light through mask. (d) Unexposed portions of photoresist dissolved. (e) SiO$_2$ layer etched where not protected by exposed photoresist. (f) Exposed photoresist removed.

Each wafer is then placed on a vacuum chuck that holds it in place by suction, and spun at high speed. **Photoresist** is applied with an eye-dropper, and is spread evenly over the surface by centrifugal force. This layer is between 0.4 and 0.7 micrometer. The thinner the layer, the sharper the engraving that takes place in the next part of the process will be, but it must not be too thin or it will fail to protect those parts of the oxide layer that are not to be etched. After coating, the wafers are baked in a low-temperature oven (under 120° C) to harden the resist and increase its adherence.

This part of the process is similar to photography. The resist is sensitive to ultraviolet light, which makes it insoluble. A **photomask** with the circuit details is positioned over the wafer, and it is exposed to ultraviolet light. Where the light passes through the mask, the resist is rendered insoluble; but where it is blocked, the resist is unaffected. The wafer is then immersed in a bath that dissolves the unexposed portions of the resist, and is then given another baking in a low-temperature oven to harden the undissolved resist.

The next step is to immerse the wafer in a buffered solution of **hydrofluoric acid**. This acid etches glass and other silicates, so it eats away the oxide layer wherever it is not protected by resist. The acid opens "windows" in the oxide to expose the silicon underneath.

The wafer is now ready for the **diffusion process** that makes the source and drain regions. Earlier in the chapter you saw that a p-channel enhancement-mode MOSFET requires two p-type regions formed by diffusing boron into the n-type silicon of the substrate. This is done by placing the wafer in an oven with a temperature of about 1200° C in an atmosphere of gaseous boron. Some of the boron condenses on the surface of the silicon by *deposition*. When the concentration is high enough, the wafer goes to a second oven at a higher temperature and the boron is "driven in." This part of the process is called *distribution*. The atmosphere in the distribution oven is an oxidizing one, so a new silicon dioxide layer is also grown over the silicon in the windows. The boron penetrates the silicon to a depth of about 7 micrometers during some 16 hours.

The wafer is then removed from the oven, and a new coating of resist applied (the residue of the old one was removed before the diffusion process). A new photomask is used to form a window through which the silicon dioxide is etched away over the gate area. The remaining resist is then removed.

The next step is to form another layer of silicon dioxide on the exposed silicon in the gate region. Wait a minute! Didn't we just remove the oxide from there? We certainly did, and the reason was because it was much too thick for the gate dielectric, which should typically be 0.1 micrometer thick. This time the oven temperature and length of time of exposure to oxygen must be controlled very precisely to obtain the correct result.

We now have to deposit a layer of metal over the entire surface of the wafer. When the unwanted parts have been etched away, the remaining metal pattern will form the gate electrodes and connections between the circuit elements. Aluminum is the metal generally used because it has low resistivity and forms strong mechanical bonds with the surface oxide. With silicon it forms a polycrystalline "microalloy." This is a valuable characteristic, because it avoids creating parasitic p-n junctions, as the alloy joint is purely **ohmic**.

Before the aluminum layer is deposited, windows have to be etched in the oxide covering the source and drain. This step is no different from previous similar ones. The wafer is then placed in a vacuum chamber where the temperature is somewhere between 200° and 400° C, and the pressure is pumped down to about one millipascal. A small quantity of aluminum is also present in the chamber, and a means is provided for heating this to above its vaporization point (2477° C), electron beam, induction or resistance heating being usually employed. This vapor condenses on the wafer, which by comparison with the vaporized metal is extremely cold.

Using another photoengraving routine, the aluminum conductor pattern is formed, which completes the fabrication of the circuits on the wafer.

Bipolar Fabrication

The fabrication of bipolar devices begins, in the same way as that of MOSFETs, with the production of wafers of single-crystal silicon. The subsequent steps for a single BJT are shown in Figure 1.15.

First, an oxide layer is formed on the surface of the silicon substrate. In this a window is etched (A), and through this window n+ material is diffused to form an n+ layer (B). This layer, which is called the **buried layer**, is more conductive than a normal n-type semiconductor because it has many more free electrons due to the richer doping. Its purpose is to reduce the effective collector resistance.

The rest of the oxide layer is now removed, and an n-type layer of silicon is deposited all over the substrate, including the n+ layer, by **epitaxy** (C). This is a process in which the layer is applied by dipping the wafer into a molten solution of n-type silicon, removing it, and then allowing it to cool so that the new n-type layer crystallizes out as single-crystal silicon also.

Another method of obtaining this layer is by heating the substrate to 1200° C in an atmosphere of silicon tetrachloride and hydrogen. The hydrogen and chlorine combine, and silicon is deposited on the substrate. This method is known as **vapor epitaxy**.

A new oxide layer (D) is now formed by passing oxygen over the heated wafer, and this layer is next etched (E) to allow p-type impurities to be diffused right through the n-type layer, so as to di-

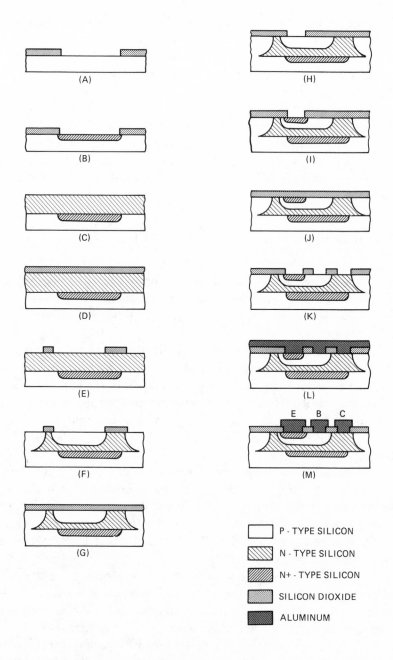

(A)

(B)

(C)

(D)

(E)

(F)

(G)

(H)

(I)

(J)

(K)

(L)

E   B   C

(M)

☐ P - TYPE SILICON

▨ N - TYPE SILICON

▨ N+ - TYPE SILICON

▧ SILICON DIOXIDE

■ ALUMINUM

Figure 1.15 Bipolar fabrication. It takes six more steps to make this npn BJT than it does to make a MOSFET.

vide it into segments (F) that are isolated from each other by reverse biasing. Each of these segments can then be an individual transistor, resistor, diode, or capacitor, as required.

A new oxide layer is formed (G), and a window is etched over the base region (H). Through this window n+ material is diffused (I) to form the emitter region. Oxide is then formed over its surface (J), and another etching is performed to make the emitter, base, and collector windows (K). A final aluminum layer is formed (L), and into it is etched the pattern for the metal interconnections (M), as in the MOS technology already described.

Ion Implantation

All of the preceding processes have used diffusion to produce their doped regions. However, in diffusion the dopant tends to diffuse sideways under the mask as well as downward as it penetrates the silicon, giving an ill-defined edge to the doped region. By using an ion accelerator, precise quantities of a single dopant can be introduced into sharply defined areas to a depth of about 1 μm. After bombardment, the device is annealed (heated and allowed to cool slowly). This permits the lattice to heal the damage caused by the passage of the ions, and also lets them settle into the most suitable sites. The process is called **ion implantation**.

Electron-Beam Machining

A limit to the smallest dimension that can be defined by using the standard masking process is imposed by the wavelength of ultraviolet light used for exposure. An optical microscope is limited in a similar way by the wavelength of visible light, so that you cannot see anything smaller than about 0.25 μm. The electron microscope is able to "see" smaller objects at greater magnification because the electrons in the beam it uses instead of light have a wavelength 100,000 times shorter than the wavelength of light.

In the **electron-beam machining process** an electron beam is made to scan the substrate in much the same way as in a TV picture tube. The beam is modulated so that electrons impinge on a special coating only where a line or area is to be masked. These portions are hardened so that when a solvent is applied they remain to form a mask that is practically integral with the substrate. Lines only 0.1 μm wide can be produced by this method.

Performance Testing

The next step is to test them. This is done by placing the wafer in an automatic testing machine that lowers a set of very tiny probes on to each circuit in turn. The test is a simple one, in which a DC voltage is applied to the proper points in each circuit, and the response is checked at the rate of several hundred tests per hour. The machine gives a "go" or "no-go" indication for each circuit, and those that fail are marked with a colored dye.

Separation, Mounting, Bonding, Packaging

Now, at last, the individual circuits can be separated. The wafer is held in a machine in which a fine diamond tip scribes grooves up and down and across between the individual circuits, so they can be snapped off in the same way in which you cut glass. Those marked with dye are discarded, as are any damaged in separation.

The final step is to package each chip. The package must provide mechanical protection, isolation from the atmosphere, and terminals for connection to other circuits. The three most popular packages, shown in Figure 1.16, are the dual-in-line (DIP), the flatpack, and the TO metal can. ICs are available in all three, although today the DIP seems to be the most common type.

Leads have to be connected between the metallized areas on the chip and the inner ends of the external terminal pins or leads of the package. This is done by thermocompression or ultrasonic bonding.

Thermocompression "nailhead" bonding is done with a machine with a bonding tip of tungsten carbide, as shown in Figure 1.17. Through this tip, gold wire with a diameter from 20 to 50 μm is continuously fed. A nozzle directs a small hydrogen flame at the projecting tip of the gold wire, which melts to form a small bead

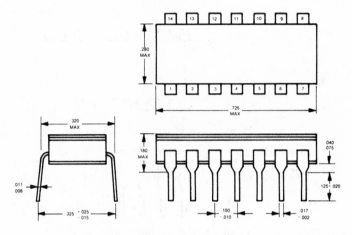

14 Lead Cavity DIP

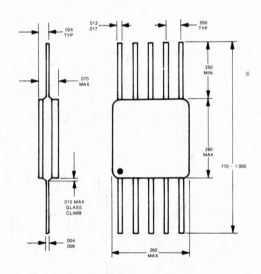

10 Lead Flat Package

Figure 1.16 Types of packaging for ICs.

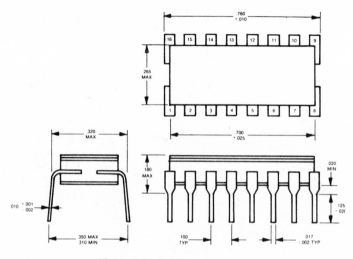

16 Lead Cavity DIP

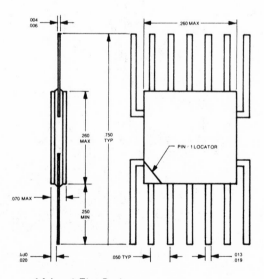

14 Lead Flat Package

Figure 1.16   (*Continued*)

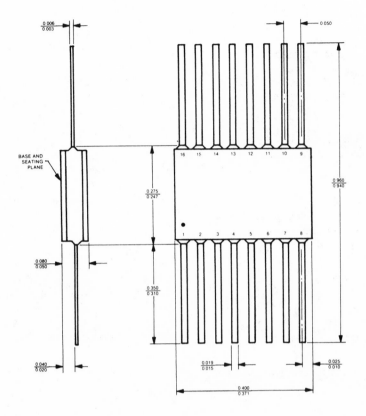

16 Lead Flat Package

Figure 1.16 (*Continued*)

(hence the name "nailhead"). The silicon or metal film to which the wire is to be bonded is positioned under the wire, and the bonding tip, which is electrically heated, is brought down on the bonding point to form a thermal-compression bond.

The bonding tip is then repositioned over the other connection point, and brought down on it. However, this time there is no ball to make a ball bond, so the machine makes a wedge bond. It also breaks the wire off, and the hydrogen flame forms a new bead at the tip of the wire, ready for the next bonding operation.

In ultrasonic bonding, ultrasonic vibration replaces the heat used in forming a thermal-compression bond. The vibration breaks up surface oxides on the metal surfaces being bonded so that the metals are brought into intimate contact and weld together.

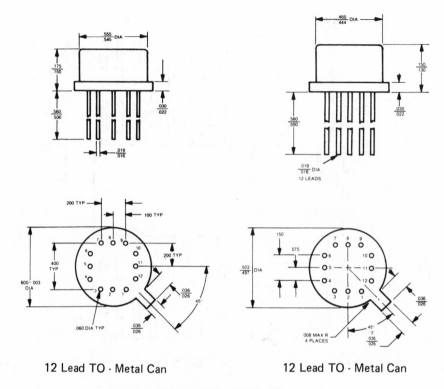

12 Lead TO - Metal Can   12 Lead TO - Metal Can

Figure 1.16 (*Continued*)

After bonding, the completed assembly is thoroughly cleaned and heated in a low-temperature vacuum furnace to drive out any gas or moisture. The base on which the chip is mounted, and to which the leads are attached, is called the header. In the final process a cap is placed over the chip and sealed to the header. This is done in an inert-gas atmosphere, such as nitrogen, so as to exclude air and moisture from the interior of the device.

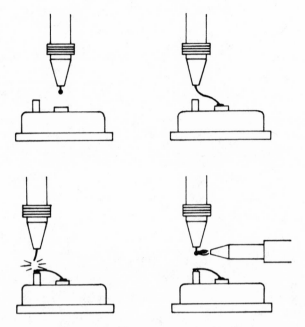

Figure 1.17 Thermocompression bonding.

CHAPTER **2**

Family Background Is Important

Digital Integrated Circuits

There are two ways of using common-emitter or common-source amplifiers: **digital** and **linear**. This and the following three chapters are concerned with digital devices.

Digital devices manipulate data (that is, information) in the form of **pulses**. A pulse of the proper amplitude will bias a transistor to saturation or cutoff, depending upon its polarity, and the polarity of the transistor. When saturated, a transistor is fully conducting. It is practically a short circuit. When cut off, its conduction is nil. Transistors specially designed for use in this way are called switching transistors.

There are four fundamental operations performed by digital circuits: NOT, AND, OR, and memory. These circuits are called **logic circuits**. Memory is discussed in Chapter 4. This chapter is concerned with the first three operations, which are termed **gates**. Standard symbols for these gates are shown in Figure 2.1.

The data is in the form of the binary digits 1 and 0.[*] A 1 is represented by a higher voltage level; a 0 by a lower, or zero level. Binary digits are abbreviated as "bits."

---

[*]See binary number system in Appendix V.

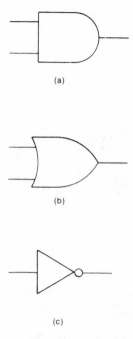

(a)

(b)

(c)

Figure 2.1   Three basic logic gates: (a) AND; (b) OR; (c) NOT.

NOT, AND and OR Gates

Figure 2.2 illustrates a bipolar NOT gate, or inverter. As you can see, it is a simple common-emitter amplifier. With an NPN transistor, as shown, the supply voltage is positive, and the transistor requires a positive voltage on its base to make it conduct.

Without a positive voltage on the base, the transistor-switch is off, therefore the voltage at B will be the same as the supply voltage. When a positive voltage is applied at A, it turns the transistor-switch on, so that B is now approximately at zero potential, giving a negative-pulse output. Since this is the opposite of the input, the device is an inverter.

A positive voltage at B is a 1 bit, of course. Another way of saying this is to say that the output at B is *true*. When the voltage is absent, the output is *false*. The same terms can be used for the input as well. Consequently, a table showing the operation of the switch

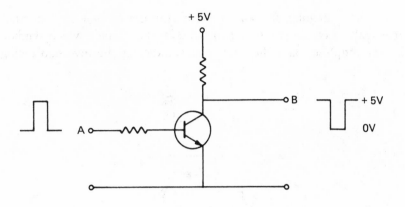

Figure 2.2 NOT Gate, or Inverter.

with different inputs is called a *truth table*. In it, the true signal is indicated by a figure 1, the false by a figure 0.

In the truth table in Figure 2.2, the false signal at A gives a true output at B, which means, in this case, that no input at A gives a +5-volt output, because the transistor is cut off. When a true signal is received at A, the voltage at B drops to zero, which is a false output. As you can see, a true input gives a false output and a false input gives a true output. Thus the device is called a NOT gate because its output is *not* the same as its input.

The triangular symbol represents the device in logic diagrams. It is really a combination of two symbols: the triangle which symbolizes an amplifier, and the circle which indicates the inversion taking place.

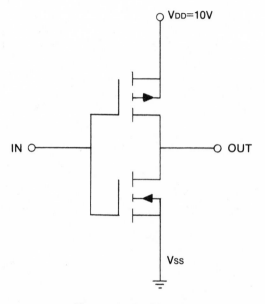

Figure 2.3   CMOS Inverter.

Figure 2.3 shows a typical CMOS inverter (CMOS is discussed on page 79). V_{DD}–V_{SS} may be anywhere from 3 to 18 volts.* In a majority of cases you would find V_{SS} at ground potential, so V_{DD} can be any value in the given range. With 0 input, the driver device is not conducting because it is an n-channel enhancement-mode MOSFET with a V_T of about 2 volts. The load device, being a p-channel device, does conduct because the gate voltage is negative with respect to V_{DD}. This connects V_{DD} to the output, so the 10-volt supply is dropped across the driver device, and the output is therefore a 1 bit. When the input is 1 ("high"), the driver device conducts and the load device turns off. This removes the V_{DD} voltage from the output, which is now connected to ground, so the output is 0. Thus a 0-bit input gives a 1-bit output, and a 1-bit input gives a 0-bit output.

---

*This makes CMOS devices very flexible, a decided advantage.

In figuring out schematics of gates using MOSFETs, all you have to remember is that a 1 bit turns n channels *on* and p channels *off*, while a 0 bit does just the opposite. Also, n channels are those with the arrow pointing toward the gate, p channels have the arrow pointing away from the gate.

The gates of CMOS devices are their weak point, and it is usual to provide them with protective circuitry on the chip. Figure 2.4 is a typical example. The breakdown voltages of the diodes are well within the 70 to 100 volts breakdown potential of the gate oxide. The RC time constant of the circuit is too low to have any appreciable effect on circuit speed.

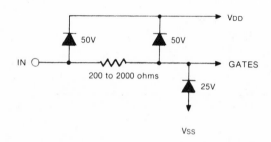

Figure 2.4 Gate protection circuit.

Figure 2.5 shows an AND gate. The transistors Q1 and Q2 are connected in series, so that both have to turn on before current can flow through them. The collector of Q1 remains at the supply voltage until true signals are received at both A and B, when it drops to approximately zero. This turns Q3 off, and the voltage at C rises to the value of the supply voltage. The effect of different inputs is shown in the truth table.

Figure 2.6 shows an OR gate, which is activated by a true input at either A or B, since the transistors are in parallel. When both transistors are off, the voltage at C will be zero. When either or both turn on, current flowing through R2 causes the voltage at C to rise to approximately the value of the supply voltage. Consequently, a true input at either A or B gives a true output at C, as shown in the truth table, which is part of Figure 2.6.

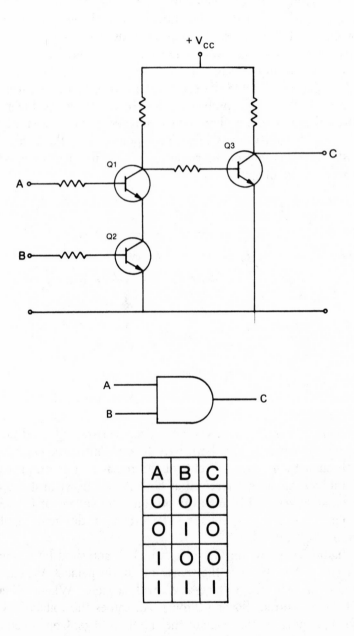

Figure 2.5 AND Gate.

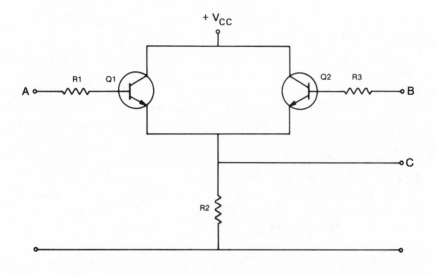

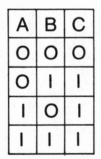

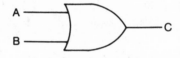

Figure 2.6 OR Gate.

NAND and NOR Gates

The truth tables for the OR and AND gates, tell you that with true inputs you get true outputs. However, for purely practical reasons, the manufacturers of integrated circuits prefer gates where false inputs give true outputs, and vice versa, so circuits in which inversion takes place are used. An inverter is a NOT gate, so an OR gate with inversion is called a NOR gate (NOT + OR), while an AND gate with inversion is called a NAND gate (NOT + AND). Rather than use OR and AND gates, the present practice is to add inverters to NOR and NAND gates when it is necessary to have an output of the same sign as the input.

The OR gate of Figure 2.6 is easily changed to a NOR gate by transferring R2 to the collector circuit, and taking the output C from the collectors of the transistor switches Q1 and Q2 instead of from the emitters as in Figure 2.11. Similarly, the AND gate of Figure 2.5 could become a NAND gate by the removal of Q3. The output signal from the collector of Q1 would then become the output signal of the gate, and would be inverted, of course.

The symbols for these gates are shown in Figure 2.7(a). Note that the NOR and NAND logic symbols are distinguished from the OR and AND symbols by the addition of a small circle at the output. This indicates that the output will be the opposite of that of OR and AND gates, as shown in the truth tables. We shall describe some standard NOR and NAND circuits later in the chapter.

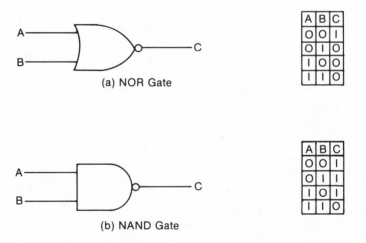

(a) NOR Gate

A	B	C
O	O	I
O	I	O
I	O	O
I	I	O

(b) NAND Gate

A	B	C
O	O	I
O	I	I
I	O	I
I	I	O

Figure 2.7(a)   NOR and NAND Gates.

Exclusive OR and NOR Gates

The exclusive OR gate's output is true if either input is true, but *not* if both are true; the exclusive NOR gate is the inverse of the exclusive OR gate. The symbols for these gates are shown in Figure 2.7(b).

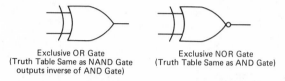

Exclusive OR Gate
(Truth Table Same as NAND Gate
outputs inverse of AND Gate)

Exclusive NOR Gate
(Truth Table Same as AND Gate)

Figure 2.7(b) Exclusive OR and NOR Gates.

Threshold Voltage

In Figure 2.8 two inverters are shown directly connected. When a false signal is applied at A, Q1 is turned off, and the voltage at B rises toward the supply voltage. As B is connected to the base of Q2, the voltage must be rising there also. At about 0.7 volt, Q2 begins to conduct, and at about 0.9 volt, Q1's collector voltage levels off and does not rise any more. Q2 is now fully conducting and the

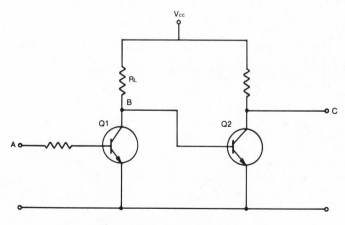

Figure 2.8 Inverters cascaded.

0.9 volt is dropped across the emitter-base junction. The remaining V_{CC} potential is dropped across R_L.

The voltage (0.7 volt) at which Q2 turns on is its **threshold voltage**. This is defined as that voltage at the input of a circuit when it changes its state. It depends upon the type of semiconductor in use, and upon the architecture of the device. As we've seen, a silicon junction transistor's threshold is about 0.7 volt, but that of a MOSFET is usually around 2 volts.

Noise Margin

The threshold voltage is important in determining the **noise margin**. Noise is defined as random unwanted signals which tend to interfere with proper perception of the desired signals. In logic circuits, noise is any unwanted voltage appearing at the input of a gate which might cause it to change state when it shouldn't. There are really two noise margins. If the input is a logical 0, a noise signal that exceeds the threshold voltage will cause the device to change its state. On the other hand, if the input is at a logical 1, a noise signal that decreases the input voltage level to less than the threshold voltage will also cause the device to change its state. This is illustrated in Figure 2.9.

In specifications, the noise margin given for the circuit is the

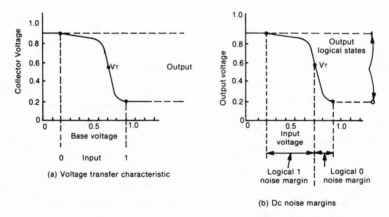

Figure 2.9 Direct-coupled circuit characteristics.

one with the lower value. As far as possible, devices are designed to have roughly equal noise margins above and below the threshold voltage. In the example in Figure 2.9, the circuit noise margin will be 0.2 volt, which makes it very vulnerable to noise.

Propagation Delay

The **propagation delay** of a circuit is the time interval between the application of a signal to the input and the appearance of the consequent voltage change at the output. In integrated circuits the interconnections are so short that we can ignore current propagation time along the conductors (about 23 centimeters per nanosecond!). The delay is caused by the charging or discharging of the capacitances in the transistors forming the gates, together with the stray capacitances and circuit resistances. Where gates are cascaded, the total delay time will be the delay per gate multiplied by the number of gates. Different types of circuits, therefore, have different operating speeds, as we shall see. Propagation delay is also called **gate delay**.

Fan-In and Fan-Out

Logic gates have input and output impedances just like other circuits, and these determine what can be connected to them. If the impedance of the input permits the connection of more than one preceding circuit, it is said to have a **fan-in** of whatever is the maximum number that can be connected. Similarly, the maximum number of following circuits that can be connected to the output is the **fan-out**. Since the input impedances of these circuits are in parallel, the driving gate's internal output impedance must be low enough to allow the full logic-voltage swing to be produced across the low impedance shunted across it.

Logic Families

So far, in this chapter we have discussed the logic functions and how they might be implemented by MOS and bipolar devices.

Several logic-circuit families or groups have evolved, distinguished by the method of carrying out the logic function and the manner of coupling between stages. They are designated by abbreviations of their full names, and these can be confusing at first. However, remember that any abbreviation containing the letters TL refers to bipolar transistor logic, whereas MOS devices all contain the letters MOS.

DCTL

DCTL (Direct-Coupled Transistor Logic), as its name tells us, is a type of circuit in which the output of one gate is connected directly to the input of the next. A DCTL NOR gate is shown in Figure 2.10. Input A is connected directly to the collector of the previous stage (B and C are connected similarly, but their previous stages are omitted since one example will suffice). When any input is high (1), the corresponding transistor turns on, virtually grounding its output to give a logical 0 bit.

The advantage of this circuit is its simplicity, but it has a serious problem also. Since it is almost impossible to get three (or what-

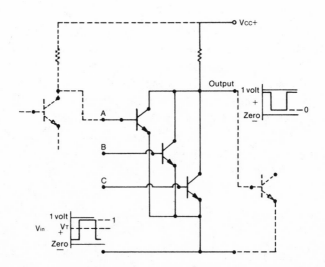

Figure 2.10 Basic DCTL NOR circuit.

ever number) of identical transistors in the IC, one will always have a lower base-emitter potential than the others, and will therefore take more than its proper share of the available current, which may prevent proper overall operation of the circuit. This characteristic is known as "current hogging."

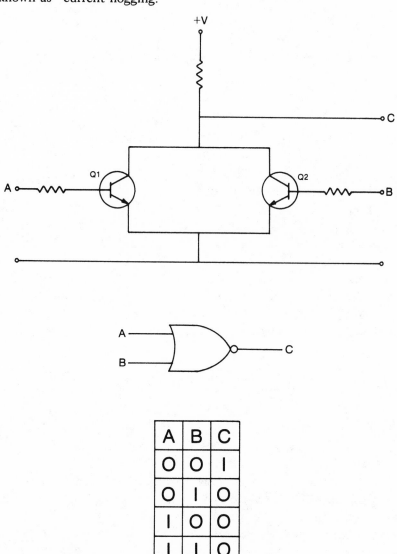

Figure 2.11 RTL NOR Gate.

RTL

RTL (Resistor-Transistor Logic) circuits have a resistor in series with each base lead, as shown in Figure 2.11. These resistors reduce current hogging but increase circuit delay, since they form RC combinations with the circuit capacitances (such as junction capacitances). Each RC element has a definite charging time that increases propagation delay. There is also a noise margin problem.

DTL

The DTL (Diode-Transistor Logic) circuit performs the NAND function. An example of this circuit is shown in Figure 2.12.

As long as either diode can conduct, Q1's base voltage is low, and Q1 is cut off. As Q1's emitter is at zero potential (with no current flowing through Q1), Q2 is also cut off. When true signals appear at A *and* B, both diodes are reverse-biased and cannot conduct. Consequently, Q1's base voltage rises, turning Q1 on. The voltage drop across Q1's emitter resistor turns Q2 on, resulting in a false output at C.

The DTL circuit is not particularly fast, but it produces a large logic swing and has a good noise margin, so it is suitable for medium-speed switching operations.

TTL

The TTL or T²L (Transistor-Transistor Logic) NAND gate is illustrated in Figure 2.13. This is a very widely used gate, with a higher speed of operation than the DTL. It works in much the same way.

Q1 is a special transistor with two emitters that perform the same function as the DTL input diodes. Actually, you can have up to ten emitters with this type of transistor just by etching additional windows in the oxide for emitter diffusion; consequently it is termed a **multiple-emitter** transistor. When either or both emitters have a false input, the output from Q1's collector is essentially zero. As a result, Q2 is cut off. The voltage on Q2's emitter is zero, that on its

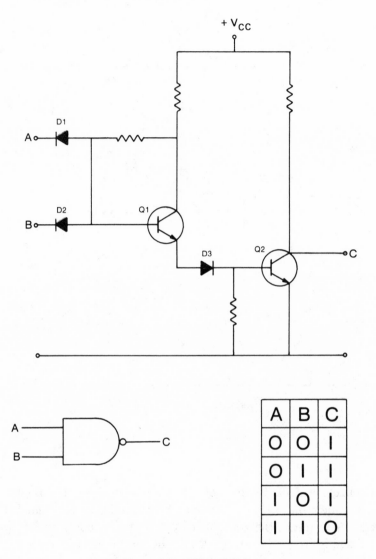

Figure 2.12 DTL NAND Gate.

A	B	C
O	O	I
O	I	I
I	O	I
I	I	O

collector the same as the supply voltage. In other words, with a false signal on its base, Q2 has two outputs: false from the emitter, true from the collector. Q2 is, therefore, a phase-splitter.

Q3 and Q4 are connected in series (a "totem pole"). When

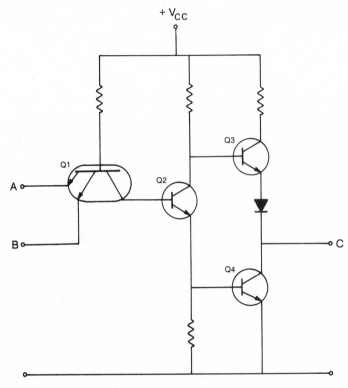

Figure 2.13   TTL NAND Gate.

Q2's collector is true, the base of Q3 is true also, so Q3 is turned on. At the same time, Q4 is turned off by the false signal from Q2's emitter. As a result, the collector of Q4 is at the supply voltage (all the voltage is dropped across Q4), and the output at C is true.

When true signals are applied simultaneously at A and B, a positive (true) voltage appears on the base of Q2, so its two output signals are now the opposite of what they were before. Consequently, Q3 turns off and Q4 turns on. Now a false (zero) output appears at C.

Although TTL has a higher speed of operation than DTL, it may, if not carefully designed, be guilty of current hogging. Because of variations in the inputs, current may be shunted away from one or more of the inputs into the others.

CML (ECL)

The CML (Current-Mode Logic) or ECL (Emitter-Coupled Logic) gate is much faster than either of the two previous devices. The example in Figure 2.14 has three inputs, applied to the bases of Q1, Q2 and Q3. The emitters of these transistors are coupled to the emitter of Q4, and all four share a common emitter resistor R2. This resistor's value is high enough so that it can act as a constant-current source.

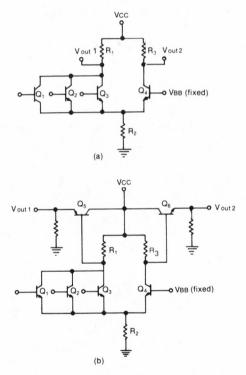

Figure 2.14 CML (or ECL) OR/NOR Gate: (a) the basic circuit; (b) the circuit incorporating emitter follower outputs.

The base of Q4 is connected to a fixed voltage, V_{BB}, that is the same as the threshold voltage for this transistor. The value of R3 is such that the voltage drop across it—when all the circuit current is flowing through Q4—is equal to the difference between the true and false voltage levels. This results in a false output at $V_{out}2$ and a true output at $V_{out}1$.

If, however, one of the inputs receives a true signal, its associated transistor is turned on. Since the voltage on its base is higher than that on the base of Q4, it conducts more heavily, and so a higher share of the current passing through R2 now goes through R1 than goes through R3. This causes the output at $V_{out}1$ to change from true to false, and that at $V_{out}2$ to change from false to true.

Since a NOR circuit is usually required more than an OR circuit, the output at $V_{out}1$ is used rather than that at $V_{out}2$. Also, in order to improve the output signals, both outputs are applied to the bases of emitter-coupled transistors Q5 and Q6. These have very low output impedances, allowing for a fan-out as high as 25 each.

I²L

I²L (Integrated Injection Logic) is a refinement of bipolar logic that eliminates resistors altogether. The circuit in Figure 2.15 is that of a NOR gate that works in much the same way as the RTL gate shown in Figure 2.11. However, as you can see, the input resistors have been replaced by Q1 and Q4, which provide the drive currents for the bases of Q2 and Q3. Since Q1 and Q4 are PNP transistors,

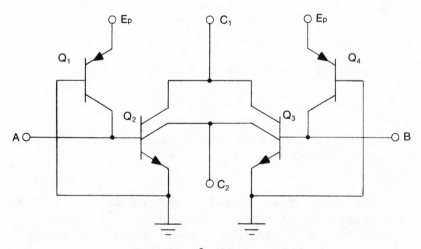

Figure 2.15 I²L NOR gate circuit.

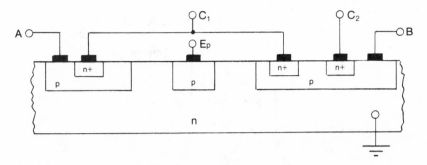

Figure 2.16 I²L NOR gate construction.

they inject an excess of minority carriers into the NPN transistors Q2 and Q3. The latter may be multiple-collector transistors, allowing for many different circuit designs.

The most significant feature of I²L circuits is the way in which the transistors are merged on the chip. The circuit in Figure 2.15 will actually be fabricated as in Figure 2.16, and as it only requires five fabrication steps it can be made much more cheaply than regular bipolar circuits or even MOS. Device density is also high: one manufacturer's single-chip microprocessor contains 6330 logic gates with propagation speeds approaching those of conventional T²L logic.

STL

STL (Schottky Transistor Logic) makes use of the extremely high switching speed of Schottky barrier diodes (semiconductor/ metal junction diodes) in circuits constructed similarly to I²L. The combination of the high density of I²L with a gate delay of 1 ns makes this technology very attractive for VLSI (very large-scale integration) arrays, such as microprocessors. Figure 2.17 shows an STL inverter with a fan-out of three outputs. Apart from their speed, these diodes operate in the same way as ordinary PN diodes. They are usually constructed of n-type silicon and gold.

PMOS

This is the simplest MOS process, consisting of a p-channel enhancement-mode device fabricated on (111) plane silicon.

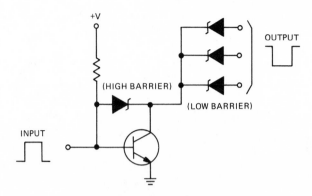

Figure 2.17   STL Inverter.

The basic MOSFET IC is inferior to the bipolar IC in its **speed/power product**. This is a figure of merit derived by multiplying the frequency response by the power consumption. Although the power consumption of MOS circuits is lower, the faster switching speeds of junction transistors raise their speed/power figure by at least an order of magnitude over that of the MOSFET.

Another difficulty arises when a MOS-bipolar interface is required. The threshold voltage of the basic MOSFET is around -4 V. For interfacing without the addition of level shifters, circuit buffering and auxiliary power supplies need a threshold voltage between -1.5 V and -2.0 V.

On the other hand, MOSFETs are smaller, so that more complex circuitry can be fabricated on a chip of given size than would be possible with bipolar devices. Also, they are more economical to manufacture, requiring fewer steps in the fabrication process. Other advantages include a very high input impedance ($> 10^{14} \Omega$), the normally *off* condition of the enhancement-mode MOSFET, and self-isolation. In bipolar devices, a space-consuming extra diffusion has to be provided for this purpose.

These by no means end the list of plus factors for MOSFETs, but serve to show why a strong motivation existed for further development to overcome their less favorable aspects.

In discussing MOS devices, you will need to know about another parameter of MOS ICs called **field threshold voltage** (V_{TF}). The metal interconnections on the chip are deposited on an SiO_2 layer considerably thicker (≈ 1.35 μm) than the gate oxide. Nevertheless, if the voltage on the metal interconnections is high enough, inversion layers will be created in the silicon substrate beneath, which will cause short circuits. The V_{TF} for standard (111) plane MOS circuitry is between -30 V and -50 V, so the designer must ensure that his power supplies are not only lower than this, but that they are sufficiently lower, so that noise and transients cannot raise them to any level where they can bring about a breakdown in isolation.

In addition to carrier mobility which, as we have seen, is slower in the surface layer of a MOSFET than in the three-dimensional silicon of a bipolar transistor, we must also deal with the effects of threshold voltage, capacitance and width-to-length ratio (W/L) of the channel (see Figure 1.1 again).

In the basic (111) **plane process**, the value of V_T is between -3 V and -5 V. We calculate the actual power-supply requirements for this type of MOS IC as follows:

$$\text{Drain supply } (V_{DD}) \approx 3 \times V_T; 3 \times -4 = -12 \text{ V}$$

$$\text{Gate supply } (V_{GG}) \approx 6 \times V_T; 6 \times -4 = -24 \text{ V}$$

These values, unfortunately, are not compatible with bipolar circuits such as TTL.

Using (100) **plane silicon** for the substrate lowers V_T to a value between -1.5 and -2.2 V. Unfortunately, it also lowers V_{TF}. This can be corrected to some extent by making the SiO_2 layer thicker (not the gate layer, but the insulating layer covering the rest of the device), but you still get a figure between -10 and -18 V. The power supply requirements for this process are:

$$\text{Drain supply } (V_{DD}) \approx 3 \times V_T; 3 \times -2 = -6 \text{ V}$$

$$\text{Gate supply } (V_{GG}) \approx 6 \times V_T; 6 \times -2 = -12 \text{ V}$$

You'll notice at once how close V_{GG} is to V_{TF}. This does not give a safe margin for noise in the power supply.

Another problem is that hole mobility in the (100) plane is lower, as we have seen. This can be improved by making the devices physically larger, but all we are doing is trading one disadvantage for another.

The **silicon nitride gate process** on (111) plane silicon substrate is a way to use the higher mobility of the (111) plane and still get a lower V_T. The dielectric constant of SiO_2 is 4.0, but that of Si_3N_4 is 7.5. In this process a layer of SiO_2 is deposited over the channel, but it is only some 20 nm thick. A layer of Si_3N_4 is deposited over the SiO_2 layer, and their combined dielectric constant is 6.8. This raises the capacitance of the gate sufficiently to lower V_T to between -1.9 and -2.9 V. V_{TF} is the same as for the standard MOSFET (-30 V to -50 V). Power supply requirements are:

Drain supply $(V_{DD}) \approx 3 \times V_T; 3 \times -2.5 = -7.5$ V

Gate supply $(V_{GG}) \approx 6 \times V_T; 6 \times -2.5 = -15$ V

This process confers several advantages. The V_T is compatible with TTL devices and V_{TF} is high. The increased dielectric constant results in improved transconductance so that the MOSFETs can be made smaller. This leads to a higher packing density with greater economy. The additional layer of Si_3N_4 makes for a better seal, which increases reliability. However, the process requires extra steps in fabrication, which detracts from the saving due to the higher packing density.

The **silicon-gate process** substitutes polycrystalline silicon for aluminum as the electrode of the gate. This silicon is doped with boron to make it a fairly good conductor. Used with SiO_2 as the dielectric, with a thickness of 0.1 μm, it gives a V_T between -1.5 and -3.5 V. V_{TF} is still at the satisfactory value of between -30 V and -50 V. Power supply requirements are:

Drain supply $(V_{DD}) \approx 3 \times V_T; 3 \times -2.5 = -7.5$ V

Gate supply $(V_{GG}) \approx 6 \times V_T; 6 \times -2.5 = -15.0$ V

As you can see, these requirements are much the same as for the nitride process. Carrier mobility is about the same also. However, there is one important difference in fabrication. The silicon-gate electrode is formed *before* the source and drain diffusions are made so it is used as a mask to define their spacing. This, therefore, is done with greater precision than can be achieved when making the aluminum electrode with a photomask. In the photomask process it is necessary to allow for a certain amount of overlap of the metal electrode over the source and drain regions to make sure that it completely covers the channel (otherwise there would be a gap in the inversion layer, and the device would not work). The elimination of this overlap removes some of the capacitance from the MOSFET. This means that when it is turned on and current begins to flow, less of the initial flow is needed to charge up parasitic capacitance, and so it builds up faster. The result is a faster device. The device can also be made smaller, and the photomasking step is gotten rid of, both of which are more cost-effective.

NMOS

Electron mobility in the (100) plane is approximately three times faster than hole mobility in the (111) plane. However, electrons are the charge carriers in n-channel devices such as the depletion-mode MOSFET in Figure 1.2. Depletion-mode MOSFETs are normally *on*, so they are not suitable for use in logic circuits, which require devices that are normally *off*.

Nevertheless, there are ways to make a depletion-mode device operate in the enhancement mode. One of these is **backgate bias**. By applying a bias of a few volts to the p substrate we can shift the threshold so that the n-channel operates in the enhancement mode. But an additional power supply is required to do this.

Another method is to dope the surface of the p-type substrate more heavily, which changes V_T to an enhancement-mode value, but electron mobility is seriously impaired.

The best approach is to use a p-type silicon gate electrode. This results in V_T going positive, which is what is needed to operate an n-channel device in the enhancement mode. High electron mobility is retained, with low V_T for bipolar compatibility, and that other important feature of the silicon-gate transistor, the self-aligned gate, is also obtained.

CMOS

By combining p-channel and n-channel devices on the same substrate, as in Figure 2.18 (a), we get **complementary MOS**

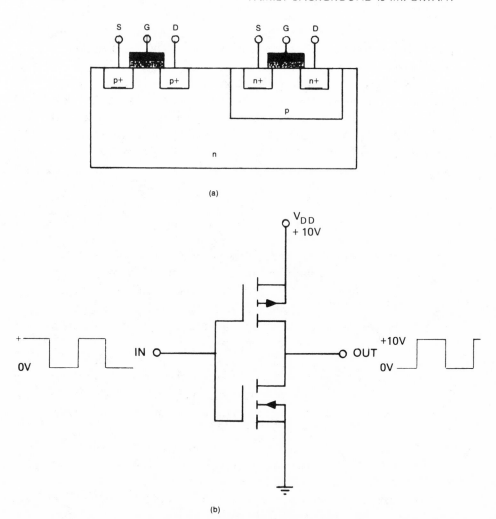

Figure 2.18   (a) CMOS structure; (b) CMOS circuit.

(CMOS, C/MOS or COSMOS). The circuit shown in (b) is operated so that one transistor is always *on*. Only one power supply is required, and practically no power is dissipated except for the brief
moment when they switch.

As you can see from the idealized waveforms, when V_{in} is positive, the n-channel device is biased *on* and the p-channel device *off.*
All the supply voltage (V_{DD}) is dropped across the p-channel device,
so V_{out} is zero. When V_{in} is negative, the p-channel device turns on

and the n-channel device turns off. Now the supply voltage is dropped across the n-channel device, so V_{out} is $+10$ V.

CMOS devices are much faster than other MOS devices, but take up more room on the chip. For this reason they would not be used in highly complex circuits such as microprocessors and large memories. However, their low power consumption makes them ideal for digital watches and other low-power applications. They are also widely used in logic circuits. Their greatly improved performance outweighs the extra cost incurred by their larger size and additional manufacturing steps.

SOS

SOS (silicon on sapphire, or spinel) provides the ultimate in speed for MOS devices by **hetero-epitaxial** growth of silicon islands on an insulator, as in Figure 2.19. Epitaxial growth means growing a crystal layer upon another crystal of a different material so that the new crystal has the same crystalline structure as the one on which it is grown. Since the silicon must be single-crystal, not polycrystalline, the insulator must be single-crystal also, and of the same type of crystal as silicon. Sapphire and spinel fit these requirements.

In an SOS device, capacitance is greatly reduced because the metal interconnections between its elements are not deposited on SiO_2 with silicon underneath, but on an insulator, so they cannot form capacitors as in a MOS device. Also PN junctions between the source or drain and the substrate are eliminated, which removes that type of capacitance as well.

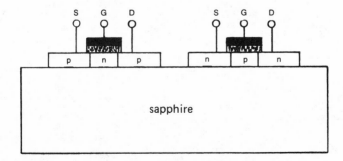

Figure 2.19 SOS structure: two MOSFET "islands" on a sapphire substrate.

The process is costly compared with standard silicon substrate fabrication, and is not used except where very high speed in a MOSFET device is desired, or where radiation hazard exists. In the latter case, the use of SOS in combination with doping the gate oxide with chromium renders MOS structures less susceptible to damage from the effects of nuclear radiation, an important feature in devices designed for military purposes.

Another good feature of SOS devices is that the isolation between elements resulting from the use of an insulator for the substrate allows for a much increased packing density, which helps to offset the additional manufacturing costs incurred by the employment of more expensive materials.

MOSFET v. BJT

Although most of the circuits in this chapter show BJTs as the active devices, MOSFETs can be and are used in nearly all. Unless extraordinary high speed of operation is essential, MOS devices will perform as well, and are cheaper to produce. Table 2.1 compares various logic families, and the principal advantages of MOS and bipolar ICs are listed on the following page.

Family	Logic Type	Propagation time per gate (ns)	Power Dissipation per gate (mW)	Noise Margin (V)	Typical Fan-in	Maximum Fan-out
DCTL (BJT)	NOR	15	10	0.2	3	3
RTL (BJT)	NOR	50	10	0.2	3	4
DTL (BJT)	NAND	25	15	0.7	8	8
TTL (BJT)	NAND	10	20	1.0	8	12
ECL (BJT)	OR/NOR	2	50	0.4	5	25
MOS	NOR	250	<1	2.5	10	5

Table 2.1 Comparison of Integrated-Circuit Logic Families.

Advantages of MOS ICs

Fewer fabrication steps
Self-isolating elements
High input impedance
Low power dissipation
High device intensity
Fewer interconnections

Advantages of Bipolar ICs

Fast switching speed
High current handling capability
High voltage gain
Lower supply voltage
No gate to break down

CHAPTER **3**

The Data Goes
Round and Around

Bistable Multivibrator

Whereas logic gates are circuits that manipulate data at high speed, a bistable multivibrator is used as a "parking place" for binary digits. It functions like a simple two-position light switch. In one position, or "state," it stores a 1 bit; in the other it stores a 0 bit.

Figure 3.1 shows the basic circuit of a bistable multivibrator. It consists of two inverters cross-coupled so that the output of each transistor is connected to the base of the other. This means that the output of each inverter is in the same state as the input of the other. The state of output Q will be the same as input B, and vice versa. To change states it is merely necessary to change the state of one of the inputs momentarily.

This type of circuit is commonly called a **flip-flop**. However, it is more precise to reserve this term for bistable multivibrators that are synchronized by a clock. A bistable multivibrator that is not clocked (that is to say, an asynchronous flip-flop) is called a **latch**. Different types of flip-flop can be used in either manner. Three commonly used types are the R-S, D, and J-K flip-flops.

R-S Flip-Flop

A basic R-S flip-flop would consist of two inverters connected as in Figure 3.2. The inputs are designated R and S instead of A and

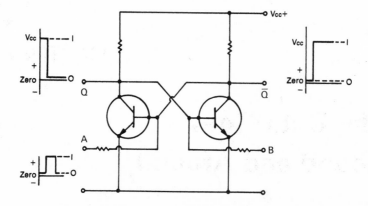

Figure 3.1   Basic bistable multivibrator.

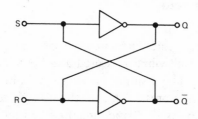

Figure 3.2   R-S Flip-Flop with Inverters.

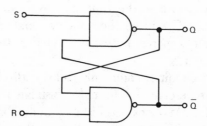

Figure 3.3   R-S Flip-Flop with Two NAND Gates.

B because they "reset" and "set" the output states. Figure 3.3 shows an R-S flip-flop using NAND gates. Since there is no provision for a clock input, this circuit could only be used as a latch. However, the latch could be gated by adding two more NAND gates, as in Figure 3.4. The circuit then ignores signals on the R and S inputs unless an

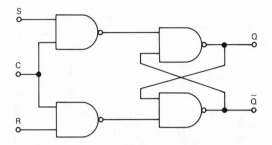

Figure 3.4 R-S Flip-Flop with Four NAND Gates.

enable signal is also present on the E input. The enable signal could be a clock pulse, of course.

D Flip-Flop

Figure 3.5 shows a D flip-flop. There are two inputs, D and C. D is for data; C is for clock pulses. The NAND gate can only accept data when a clock pulse is present on the other input. The timing diagram in Figure 3.6 shows how this works. The D line consists of a series of data pulses applied to the D input. The C line shows the clock pulses applied to the C input. The Q line shows the output state. In this flip-flop the change of state takes place on the *leading edge* of each clock pulse (in some other flip-flops this may be on the trailing edge). Q changes according to the data that is on the D in-

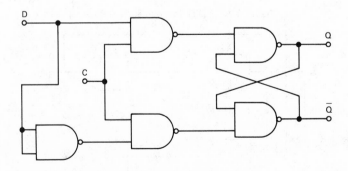

Figure 3.5 D Flip-Flop.

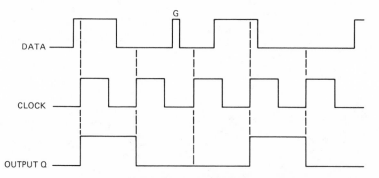

Figure 3.6 Timing Diagram for Leading-Edge Triggered D Flip-Flop. (Note how edge-triggering discriminates against random glitches such as that marked G.)

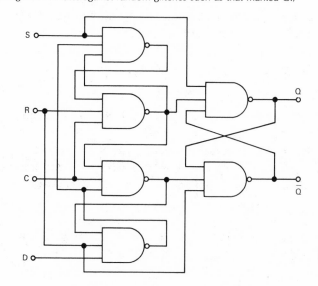

Figure 3.7 D Flip-Flop with Set and Reset Inputs.

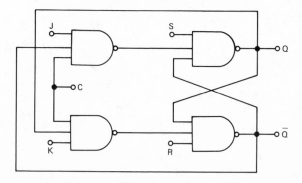

Figure 3.8 J-K Flip-Flop.

put at the moment that the leading edge of a clock pulse arrives at C, but is unaffected by what is on D at all other times. A D flip-flop can also have set and reset (clear) inputs, as shown in Figure 3.7. Pulses at these inputs can set the condition of the flip-flop at any time, regardless of clock pulses.

J-K Flip-Flop

The D flip-flop and the J-K flip-flop are the most widely used types of flip-flop. The main difference between them is that the J-K inputs replace the D input, and three-input NAND gates are therefore required, as shown in Figure 3.8. The J and K inputs may be used to control the states of the outputs, but if both are made high the outputs will toggle with the clock pulses. In addition to the J and K inputs, there are also S and R terminals for set and reset.

J-K Master-Slave Flip-Flop

An improved version of the J-K flip-flop is illustrated in Figure 3.9. It consists of eight gates and an inverter. Gates C, D, G, and H are connected to form two R-S flip-flops. They are clock-synchronized by the addition of the NAND gates A, B, E, and F. One of the inputs of each of these is tied to the common clock line from input C. This means they cannot change state until a positive clock pulse is applied.

As you can see, there are two halves to this IC. That to the left (A, B, C, D) is the *master* flip-flop; the one to the right is the *slave* flip-flop. Whatever is at the J and K inputs is transferred to the master flip-flop by the leading edge of the clock pulse at C. On the trail-

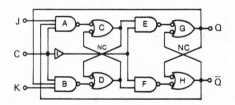

Figure 3.9 A JK flip-flop consists of a "master" and "slave." Supply voltage and reset connections are not shown (NC = no connection).

ing edge of the clock pulse, the data in the master is passed to the slave flip-flop, and is reflected in the outputs Q and $\overline{Q}$.*

If both the J and K inputs are low, the flip-flop does not change state on the arrival of a clock pulse. If the J input is high and the K input is low when the clock pulse arrives, Q will go to a 1, and $\overline{Q}$ will go to 0. If the J input is low and the K input is high when the clock pulse arrives, Q will go to a 0, and $\overline{Q}$ will go to a 1. If J and K are both high, the flip-flop will toggle (the outputs will reverse) on each clock pulse. The flip-flop cannot change state except on the leading and trailing edges of the clock pulses (it is said to be "edge-triggered"), so the chance of unauthorized changes due to random pulses or noise is minimized.

Clock

Clocks are crystal-controlled pulse generators that supply synchronizing signals to all pulse-operated circuits in a system. These clock pulses gate, or "enable," the operation of these circuits to keep them in step. The speed at which information can be handled depends upon the frequency of the clock. However, its frequency is limited by the speed of response of the devices being clocked. Where the data has to pass through several devices, the total propagation delay is the sum of the individual delays. Therefore, the clock pulses must be timed so that their intervals are more than this time, which may also vary from one unit to another, and with temperature and other conditions.

A system may also use more than one clock, even clocks running at different frequencies. Usually, the subordinate clocks are synchronized to the system, or master clock.

Clock pulses must be wide enough to fully enable the circuit being timed. However, it is advantageous to keep them as narrow as possible, since this allows the circuit to settle between pulses and also discriminates against noise, which can only be transmitted during the pulse.

---

*The bar over the Q means that the $\overline{Q}$ output is the reverse of the Q output. Pronounce it "not Q," "negated Q," "inverted Q," or "Q's complement," whichever you prefer.

Shift Register

A shift register is a *temporary* parking place for data, as opposed to a memory, which has no particular time limit. A basic shift register is shown in Figure 3.10. It consists of four D flip-flops in an IC with data, clock and reset inputs and four data outputs (V_{cc} and ground connections are not shown).

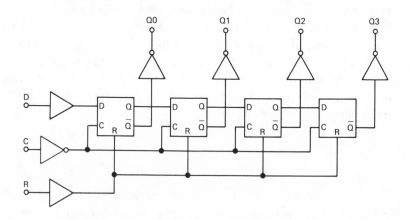

Figure 3.10 Four-Bit Serial-In/Parallel-Out Shift Register.

Data arriving at the data input, where two inverters provide buffering, are applied to the D input of the first flip-flop. This flip-flop is triggered by the leading edge of the clock pulse, and transfers the data to its Q output, and therefore to the D input of the next flip-flop. On the next clock pulse, this flip-flop transfers the data to the third flip-flop, and so on. When four bits of data have been received, each flip-flop will be storing one bit.

Output data are taken from the Q outputs of the flip-flops and inverted to appear in parallel at the Q0 through Q3 output terminals. Since the data are input in series but are output in parallel, this is called a serial-in/parallel-out shift register. Shift registers also are manufactured that accept data in parallel and output it in series. In these, the data bits are applied simultaneously to the D inputs of the flip-flops and are then shifted out of the final Q output by successive clock pulses.

Shift registers are available that handle 8, 16, 32, 64, and 128 bits, as well as other numbers, and some may be used for data "words" of variable length.

A *left/right* shift register does not have parallel input or output but, as its name implies, can shift the data either way according to the state of the left/right control input. It can also keep the data recirculating (from the last output to the first input) when the recirculating control input is enabled. This can be used to provide time delay where required. The left/right alternative is useful if you want to reverse the data on a last-in/first-out (LIFO) basis.

A set (typically 16) of four-bit parallel registers can be used to accept data in parallel form and shift it from the first register to the second, and so on, until it appears at the output. Since the first data input is also the first output, this is known as a first-in/first-out (FIFO) register. It is used in computers for buffering, and in some other applications, such as automatic dialing.

Counter

Counters also consist of arrays of flip-flops. In Figure 3.11(a) four J-K flip-flops are connected to count pulses arriving at the input. All the J and K inputs are permanently high, so each flip-flop toggles on pulses applied to its clock input. Its Q output is connected to the next clock input, and to a lamp.

Typically, the state of a given flip-flop changes when its clock input goes from 1 to 0 (that is, it triggers on the trailing edge of the pulse). When the first input pulse does this the state of FF1's output goes to 1, and lamp 1 lights. The lamps now read 1 0 0 0, or binary 1* (the least significant bit is on the left in this example). When the second pulse goes from 1 to 0, FF1's output goes back to 0, and its lamp goes out. Simultaneously, FF2's input also goes from 1 to 0, so FF2's output goes to 1, and its lamp lights. The lamps now read 0 1 0 0, or binary 2. On the arrival of the trailing edge of a third pulse at FF1's input, its output goes from 0 to 1, and its lamp lights, but FF2's lamp remains on as FF2's state does not change. The lamps now read 1 1 0 0, or binary 3. On the arrival of

---

*See Appendix V for Binary Number System.

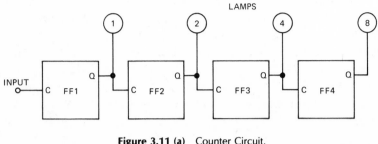

Figure 3.11 (a) Counter Circuit.

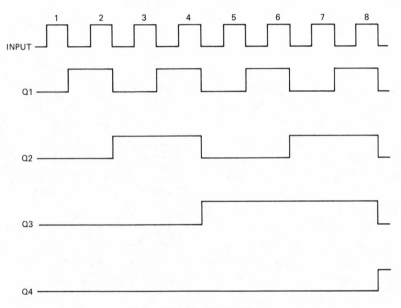

Figure 3.11 (b) Timing Diagram.

the fourth pulse, the outputs of FF1 and FF2 both go from 1 to 0, and their lamps go out, but the output of FF3 now goes from 0 to 1, so its lamp lights. The lamps now read 0 0 1 0, or binary 4. On the fifth pulse, FF1's output goes from 0 to 1, so its lamp lights, but the other flip-flops are not affected. The lamps now read 1 0 1 0, or binary 5. The sixth pulse sends FF1 back to 0, but FF2 goes to 1, so the lamps now read 0 1 1 0, or binary 6. The seventh pulse puts FF1 back up to 1, so the lamps read 1 1 1 0, or binary 7. Finally, the eighth pulse sends FF1, FF2, and FF3 to 0, turning their lamps off, but changes FF4 to 1, turning its lamp on. The lamps now read 0 0 0 1, or binary 8.

This can continue until all lamps are lit, meaning binary number 15. The next pulse after that will turn all the lamps off, taking the count back to zero, so this counter can count only 16 pulses. A **hexadecimal** count is required for a number of purposes though, so this counter, which is called an **asynchronous** or **ripple counter**, has its uses, especially where the events counted occur in a random fashion. It can also be used as a frequency divider with division factors of 2, 4, and 8, as can be seen from the timing diagram in Figure 3.11(b).

BCD Counter

However, for most purposes a decimal count (0 through 9) is required, and so the previous circuit is modified, as shown in Figure 3.12(a).

In this circuit the J and K inputs of flip-flops A, C and D do not have signal connections. They are connected to a positive voltage source, so are in a 1 condition at all times. These flip-flops will switch on the clock pulse alone. The J and K inputs of B are connected to the Q output of D. As long as this is 1, the B flip-flop will also be enabled.

When a pulse arrives at the A flip-flop's C input, the A flip-flop's Q output changes from 0 to 1 on the trailing edge of the pulse, as explained previously. This output does not trigger the B flip-flop until the next pulse into the A flip-flop switches its Q output back to 0. B then switches, and the input to flip-flop C becomes a 1. Notice that it took *two* input pulses to get *one* output pulse from A. It takes two input pulses into B (from A) to get one out of B, and so on, as you can see by looking at Figure 3.12(b), so this IC can also act as a frequency divider.

B's output to C causes its Q output to switch also, but this is not connected to D. Let's see why. We've had seven pulses applied to A's input, and flip-flops A, B and C are all in the 1 state. Now we want to get these back to 0, and D from the 0 state to a 1.

The $\overline{Q}$ output of flip-flop C is 0, and is connected to one input of NAND gate 2. The output will be a 1 for all inputs except a 1 on both. Since the output of NAND gate 1 is a 1, the output of NAND gate 2 is a 1 as long as the $\overline{Q}$ output of C is 0. When the

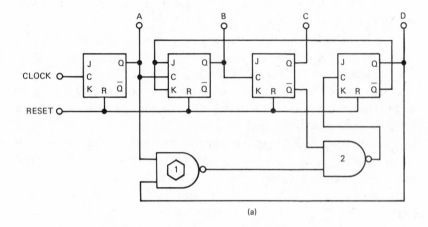

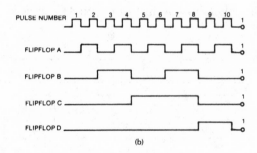

Figure 3.12 BCD Counting Unit.
(a) BCD counter using four JK flip-flops and two NAND gates. The outputs labeled A, B, C, D are connected to the read-out decoders. The R inputs are connected to a RESET push-button which enables you to reset all the flip-flops to zero at any time.
(b) Each flip-flop changes states from 1 to 0 or from 0 to 1, on arrival on the *trailing* edge of the input pulse.

other pulse arrives at A, A's state becomes 0, turning B and C off in turn. Flip-flop C's $\overline{Q}$ output is now 1, so NAND gate 2 now has two 1 inputs, and its output becomes 0. This change of D's input from 1 to 0 causes it to switch, and its Q output becomes 1. The condition of flip-flops A, B, C and D is now 0 0 0 1 (= 8). D's Q output is also applied to NAND gate 1, so when A next goes 1 it will switch and reset NAND gate 2 to its former condition of a 1 output.

On the arrival of the ninth pulse, A's output switches to 1, so both inputs of NAND gate 1 are 1, and its output is 0. This resets NAND gate 2 to a 1 output.

When the tenth pulse arrives, the A flip-flop goes to zero. This cannot affect the B flip-flop because the J and K inputs are both 0. A's Q output changes one of NAND gate 1's inputs to 0, so its output changes to 1. This puts a 1 on both inputs of NAND gate 2, so its output becomes 0. This causes D to switch to zero, so all four flip-flops have now been reset to 0.

Since each BCD counter counts from zero through nine, you can use them to count any size of number by "daisy-chaining" them, with one counter for each digit, as shown in Figure 3.13. As the tenth pulse causes flip-flop D in the units counter to return to the 0 state, it causes flip-flop A in the tens counter to change state to 1. In this way, the tens counter counts one for every ten counts of the units counter. A 1 is carried forward to the left each time the counter to its right counts to ten, just as in simple addition.

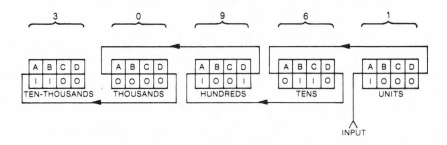

Figure 3.13 How five BCD counters are used to give a five-figure read-out.

Buffer

Buffers are generally used in conjunction with other digital devices but can be obtained in a separate IC as shown in Figure 3.14, which has six individual inverters on the same chip. The purpose of a buffer is to provide impedance separation between logic devices and amplification. Amplification is often needed for data sent over a cable between a computer and a peripheral device such as a printer, but in this case the term **line driver** may be used. Some buffers provide inversion, as in Figure 3.14, but not all. If inversion is not pro-

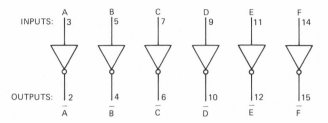

INPUTS:

| A | B | C | D | E | F |
| 3 | 5 | 7 | 9 | 11 | 14 |

OUTPUTS:

| 2 | 4 | 6 | 10 | 12 | 15 |
| A̅ | B̅ | C̅ | D̅ | E̅ | F̅ |

Figure 3.14 Hex Buffer IC. Figures denote pin numbers. Not shown are pins 1(V_{CC}), 8(GND), 13(NC), and 16(V_{DD}).

vided, the small circle will be omitted in any diagram of the IC. Other logic gates and flip-flops may also be used as buffers.

Encoders

An example of an encoder is shown in Figure 3.15. There are four OR gates, each of which requires just one true input to give a true output. The inputs come from the ten decimal inputs on the left, and produce the outputs shown in the truth table. Not shown are the connections to the shift register where the binary numbers are stored until the entry is complete, and the circuitry required to select and drive the numbers in the display.

Suppose you enter the decimal number 1906 on your pocket calculator. As you know, your calculator will display a 1 when you depress the 1 key. When you depress the 9 key the 1 is shifted to the left and the 9 takes its place. When you depress the 0 key, the 1 and the 9 both move one position to the left to make room for the 0, and lastly the 1, 9 and 0 move again to allow for the 6. This is analogous to what happens in the shift register.

However, before they reach the shift register the decimal numbers are changed to binary in the encoder. When the 1 key is pressed, one of the inputs of OR gate A becomes true, and there is a true output. The other three gates receive no input signal, so their outputs remain false. The outputs of the four gates make up the binary number 0001 as in the truth table.

The 9 key provides true inputs to gates A and D for an output of 1001, but the 0 is not connected to any of the four gates, so all

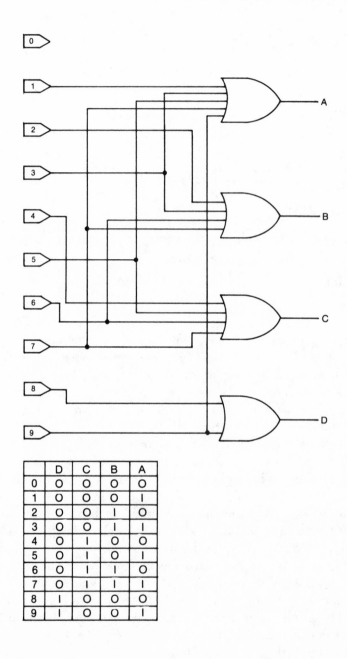

Figure 3.15   Encoder.

outputs are 0. The 6 key gives true inputs to gates B and C for the binary output 0110.

Decoders and Display Drivers*

A ROM may be used to decode binary data so it may be displayed as decimal. In the example in Figure 3.16, the binary input

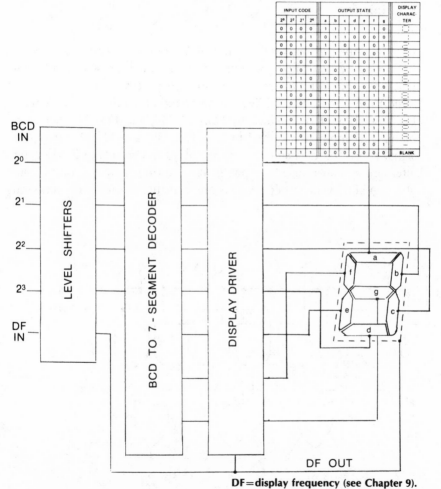

INPUT CODE				OUTPUT STATE							DISPLAY CHARACTER
2^3	2^2	2^1	2^0	a	b	c	d	e	f	g	
0	0	0	0	1	1	1	1	1	1	0	
0	0	0	1	0	1	1	0	0	0	0	
0	0	1	0	1	1	0	1	1	0	1	
0	0	1	1	1	1	1	1	0	0	1	
0	1	0	0	0	1	1	0	0	1	1	
0	1	0	1	1	0	1	1	0	1	1	
0	1	1	0	1	0	1	1	1	1	1	
0	1	1	1	1	1	1	0	0	0	0	
1	0	0	0	1	1	1	1	1	1	1	
1	0	0	1	1	1	1	1	0	1	1	
1	0	1	0	0	0	0	1	1	1	0	
1	0	1	1	0	1	1	0	1	1	1	
1	1	0	0	1	1	0	0	1	1	1	
1	1	0	1	1	1	1	0	1	1	1	
1	1	1	0	0	0	0	0	0	0	1	—
1	1	1	1	0	0	0	0	0	0	0	BLANK

DF=display frequency (see Chapter 9).

Figure 3.16 Decoder and Display Driver. As shown in the truth table, different BCD inputs result in illumination of the corresponding segments of the display.

*See Chapter 9 for various displays.

is applied at the 2^0, 2^1, 2^2, and 2^3 inputs to the level shifters. These circuits accept various voltage levels and adjust them, if necessary, to provide the proper input-voltage-level inputs to the decoder. If the binary signals were at the required level, this would not be necessary, but the manufacturer included the circuits to make the device compatible to a wide range of inputs.

Multiplexer

Multiplexers are used when signals from several sources have to be carried over a single communications channel, or when one computer has to control a number of different processes going on at the same time. Figure 3.17 shows the principle of a multiplexer. A four-bit parallel input is provided by D0, D2, D4, and D6. Another four-bit parallel input is provided by D1, D3, D5, and D7. A data-select signal consisting of a square wave enables the two sets of AND gates alternately. The four-bit output is also controlled by the output enable signal. Several other versions are available, some combined with buffers.

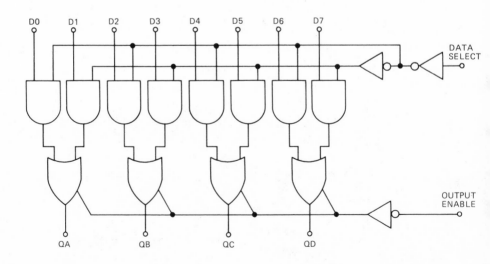

Figure 3.17 Quad 2-Input Multiplexer.

Timer

Figure 3.18 shows the circuit of the first, and still the most popular, IC timer (555). To make it work, it is necessary to connect an external timing capacitor and resistor as shown. When a negative trigger pulse is applied to the trigger input, comparator 2 resets the flip-flop to place a negative voltage on the base of Q1. This turns Q1 off, and C starts to charge toward V_{cc}. The inverting input of com-

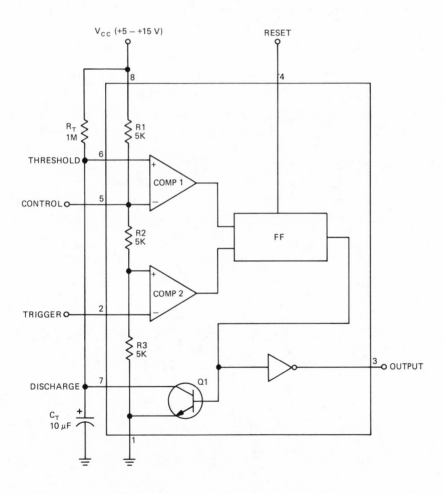

Figure 3.18 Timer (Monostable or One-Shot).

parator 1 is connected to the voltage divider (the "control voltage") between V_{cc} and ground, while the capacitor voltage is connected to the comparator's noninverting input. When the rising voltage (the "threshold voltage") on the capacitor exceeds the control voltage the flip-flop resets, placing a positive voltage on the base of Q1, so that the transistor turns on and shorts out the timing capacitor.

The inverter reverses the polarity of the output state of the flip-flop, so that a positive voltage exists at the IC output pin from the time of arrival of the trigger pulse until the moment when the threshold voltage exceeds the control voltage. The length of this pulse depends upon the values of the timing capacitor and resistor, and is given by:

$$T = 1.1RC$$

where T is in seconds, R in megohms, and C in microfarads. For instance, if R is 1 MΩ and C is 10 μF, T will be $1.1 \times 1 \times 10 = 11$ seconds.

In this mode of operation, the flip-flop is monostable; that is, it cycles once and resets on each trigger pulse. It may be made free-running by connecting pins 6 and 2 together, and connecting pin 7 to the external voltage divider, as shown in Figure 3.19. A series of

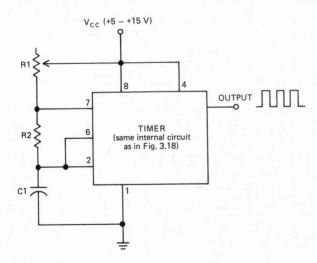

Figure 3.19 Timer (Free-Running Pulse Generator).

pulses then appears at pin 3, with a repetition rate controlled by R1. This configuration could be used for clock-pulse generation and similar purposes.

CHAPTER 4

Chips for the Memory

The two principal types of integrated-circuit memory are the **random-access memory** (RAM) and the **read-only memory** (ROM). Data is written into and read out of RAM, but generally speaking, is only read out of ROM.

Static RAM

There are two types of RAM, the static and the dynamic. The static RAM cell consists of a circuit with a flip-flop, as shown in Figure 4.1. This cell, is made up of six MOSFETs (some have eight), so it takes up a fair amount of space on a chip. The usual number of cells in a static RAM is 256 (16 × 16).

As you will recall, one of the transistors T1 and T2 will be conducting while the other is cut off. The gates of these transistors are cross-connected to each other's drains, so the one conducting holds the other off by the zero potential on its drain. Let us assume that T2 is conducting. All of V_{DD} is dropped across the load device TL2 and therefore T2's drain is at zero potential. T1 is accordingly cut off, so its drain is at the V_{DD} potential, and as this voltage is connected to T2's gate, T2 is held in a conducting state.

To read the information stored in this cell, we have to apply a negative voltage to the word select line. This voltage is applied to the gates of T3 and T4. T3 cannot turn on because its drain is at the high negative potential of T1's drain, but T4's drain is at zero potential so it turns on, and current flows from the 0-bit line through T4 and T2. This current is sensed by the 0-bit line and is therefore read as 0.

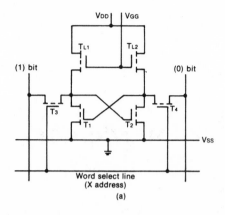

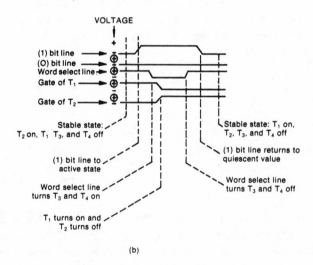

Figure 4.1 Six-transistor static RAM cell: (a) circuit schematic; (b) timing for write operation.

To write in a 1 bit on the (1) bit line, the word line is pulsed negative. This turns T3 on, so that T1's drain is grounded. This zero potential is connected to T2's gate, and T2 turns off. Its drain goes high, turning T1 on. The cell is now storing a 1 bit.

Dynamic RAM

A dynamic RAM cell uses fewer devices, so more can be fabricated on a chip. The cell in Figure 4.2 has three MOSFETs. The

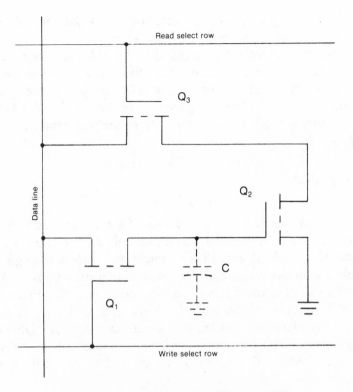

Figure 4.2 Dynamic RAM cell with three MOSFETs.

usual number of cells on this chip is 1024 (32 × 32), so you can see its storage capacity is four times as great as that of a static RAM. This makes the dynamic RAM more economical. It also dissipates less power and has a higher speed of operation.

There is no flip-flop in this cell. The data is stored by the parasitic capacitor C. To write in a bit, 1 or 0, the write select row is energized. This turns Q1 on, and C is charged according to the potential of the data line. To read the stored bit, the read select row is pulsed, which turns Q3 on. Q2 will be either on or off, depending on the stored bit, so the sense line either senses or does not sense a current.

Due to the very small capacitance of C its charge leaks away rapidly, even though the leakage current through the reverse-biased junctions is very small. To retain the data the charge must be regenerated by "refreshing." At regular intervals of a few hundred nanoseconds the cell is automatically read, and a new data bit written in before the previously stored bit can fade away.

This also protects the stored data from being erased in the readout process. Memories that retain their data regardless of how many times they are read are said to have non-destructive readout (NDRO). Most MOS memory circuits have this capability as long as they are operating, but a power supply failure will result in loss of the data.

ROM, PROM

A ROM usually consists of an array of transistors as in Figure 4.3, arranged in a rectangular pattern of rows and columns. In this example of a MOS ROM, all the sources are connected to ground, and all the drains are connected through transistor loads to V_{DD}. The outputs of the transistors in each column appear at outputs 1, 2, 3, . . . M. Input signals are applied at rows A, B, C, D, . . . N.

In manufacturing this matrix some transistors are fabricated without gates. Naturally, they won't work. If a negative voltage is applied to row A, the first transistor will do nothing as it has no gate,

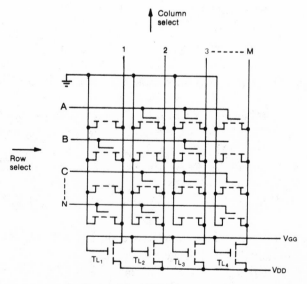

Figure 4.3 MOS ROM matrix.

so a 0 bit will be at output 1. Outputs 2, 3, 4, however, will have 1 bits because the remaining transistors in the row have gates and so will switch. The output will therefore read 0111. This constitutes a word, and row A is a **word select line**.

The pattern of active and inactive transistors comprises a permanent program that cannot be changed. You cannot store anything else in this memory, which is why it is called a read-only memory. Due to the very large number of MOSFETs that can be fabricated on a chip, it can have a very considerable storage capability.

One use for ROMs is as decoder/drivers. Binary data applied to the four input rows activate corresponding combinations of the seven output columns that energize the selected segments of a readout. ROMs are also used in calculators and microprocessors, as we shall see in Chapter 5.

A PROM is a "programmable ROM," meaning programmable by the user for a special purpose, instead of being programmed by the manufacturer by masking during its fabrication. It starts as a matrix with all gates intact, and it is programmed by destroying those required to have 0 outputs. This may be done by laser, or by providing fusible links that are melted by a high voltage or high current.

EAROM

The EAROM is an electrically alterable read-only memory. Data may be erased and new data written in up to 20 times. The most common erase mechanisms are direct electronic erase signals and application of short-wave ultraviolet or X-radiation. After erasure, the new data is routed to the specified address, and a write pulse is generated for a precise period of time to store it. Like other ROMs, the memory is nonvolatile.

Charge-Coupled Device (CCD)

This is a junctionless semiconductor device consisting of a silicon substrate covered by a layer of silicon dioxide, which in turn is capped with a row of closely spaced aluminum electrodes. Application of a potential to these electrodes results in surface depletion regions ("potential wells") beneath each electrode. The applied potential should be negative for an n-type substrate, positive for a p-type. Charge carriers of opposite polarity can then be stored in the depletion regions. If a more negative (or positive, as the case may be) pulse is applied to an electrode adjacent to one under which a charge is stored, the charge will spill over into the deeper potential well. In this way, a charge can be shifted along the chain of depletion regions. Ones and zeros are indicated by presence or absence of charge, which can be injected at the input of the chain by radiation-generated electron-hole pairs, and read out by the reverse process at the output. CCDs can provide memory, shift register, delay line, and imaging functions. (See Figure 4.4)

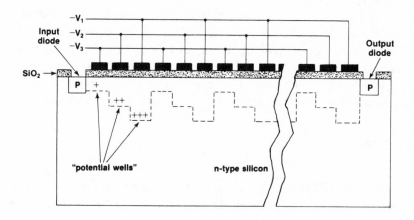

Figure 4.4   Charge-Coupled Device.

Bubble Memory

The bubble memory consists of a thin (50 μm) film of single-crystal garnet or similar magnetic material on a substrate of nonmagnetic garnet. Under the influence of a magnetic field in which the lines of force are perpendicular to its surface, very small cylindrical domains appear that are magnetized in the opposite direction to the film. These "bubbles" have a diameter of approximately 30 μm. Each one is a tiny magnet with a north and south pole.

Permalloy tracks deposited on the surface of the garnet are used to move the bubbles. An additional magnetic field, induced by two coils wrapped around the chip at right angles to each other, is used to magnetize these. This field rotates, so that the positions of the north and south poles change to conform to the lines of force turning in a plane parallel to the garnet film. The permalloy tracks are

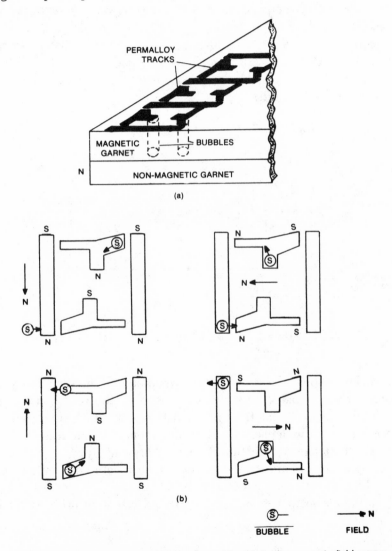

Figure 4.5 (a) Portion of a bubble-memory chip. (b) As the magnetic field rotates in a plane parallel to the Permalloy tracks, it induces magnetism in them with polarities as shown, moving the bubbles along.

shaped so that they move the bubbles along in the desired direction, as shown in Figure 4.5.

A block of data entering the input of the device generates a bubble for each 1 bit; no bubble for each 0 bit. The bubbles are transferred into a major loop (see Figure 4.6) and are lined up so that each bubble or space is opposite the end of one of an array of permalloy tracks perpendicular to the major loop. These tracks are called minor loops. The entire data block is then sent, all at once, into the minor loops. Typically, there are 256 minor loops, so the data block consists of 256 bits. Each minor loop can hold 1025 bits, so this number of data blocks can be entered, for a total storage capability of 262,400 bits.

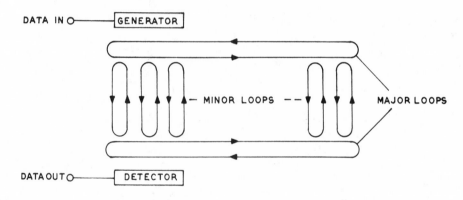

Figure 4.6 Principle of Bubble Memory (see text).

The bubbles keep circulating synchronously in their loops, so that each data block arrives in turn opposite the read major loop. Control circuitry keeps track of the data blocks, so when the desired one appears, new duplicate bubbles are produced in the read track by a transfer/replicate structure without destroying the original ones. The new bubbles are then converted to electrical signals by a detector, before being eliminated by a "bubble-eater."

Bubble memories are nonvolatile and can store up to two million bits.

CHAPTER 5

The Microprocessor Directs the Traffic

The Microprocessor

The digital devices we've covered so far can, in many cases, be used by themselves. For instance, a timer IC requires only the addition of a capacitor and a variable resistor, together with a battery, push-button switch, and LED, to be capable of being used as a dark-room exposure or developing timer. However, other equipment uses large numbers of ICs in complex systems that require some kind of automatic central control, and for these, the **microprocessor** was developed. In order to illustrate the use of this extraordinary device, we'll show how it is used in a **microcomputer**, the smaller computer used for personal, home, or small business purposes.

A Computer's IQ Is Zero

In some cities at Christmas, it is customary to leave the lights on in certain of the windows of City Hall, so that at night they form a gigantic illuminated cross. From the spectator's point of view this is very symbolic. From the point of view of the City Hall electrician inside, it means only that specified switches are in the on position and others are in the off position, in accordance with the Mayor's instructions.

Figure 5.1 shows how the basic operation of a digital computer closely parallels this Christmas story. The **user** (Mayor) gives instruc-

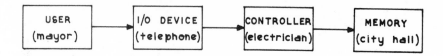

Figure 5.1   Basic Functions of a Computer.

tions via an **input/output device** (telephone) to the **controller** (electrician), who in turn manipulates the switches until the **data** (cross) are stored in **memory** (City Hall). In this sequence, the I/O device, controller, switches, and memory are all actual physical objects and are therefore called the **computer hardware**. The data, however, exist only in the form of a pattern which is termed **computer software**.

Of course, the hardware has no intelligence of its own. It is a mechanical system, operated by outside intelligence by means of a **program**, which is a sequence of instructions to the controller (which is *not* an intelligent electrician, but a **central processing unit** or **CPU**.) When a computer prints out some sequence of letters, such as "WRONG, TRY AGAIN," it is meaningful to the user, but the meaning does not arise mysteriously from within the computer; it comes from the way the programmer assigned letters to processes going on in the computer. It is really a dialogue between the user and the programmer, in which the programmer has stored all the responses required in the computer. The computer's instructions then make it select the appropriate response for any input. If computers are ever made that program themselves without human intervention it may be a different story!

The Central Processing Unit

A microprocessor duplicates the CPU of a "mainframe," or large computer, but because of its low cost, compact size and limited power consumption, it is now used extensively in all types of "intelligent" systems that include data acquisition and control, data communication, human interface (terminals, point-of-sale, etc.), computation, and the like.

The microprocessor is responsible for manipulating the data fed into the computer in accordance with its instructions. These instructions are provided in the program stored in the computer memory. The addresses of the instructions to be used are stored in proper sequence in the **program counter** (see Figure 5.2). The controller gets the address of the next instruction from the program counter, ob-

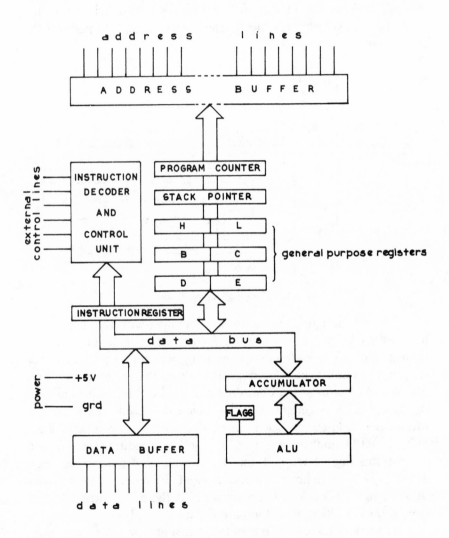

Figure 5.2 Central Processing Unit (Microprocessor).

tains the data stored in that memory address, and transfers it to the **instruction register**. This is done by means of three communication channels called the **address bus**, the **control bus**, and the **data bus** as shown in Figure 5.3. The instruction address in the program counter is placed on the address bus and readies that storage location to yield the instruction data. A signal on the control bus then enables the data to be transferred to the data bus. Another control signal fills it into the instruction register. Here it is held while the controller decodes it and issues further control signals to perform the instruction.

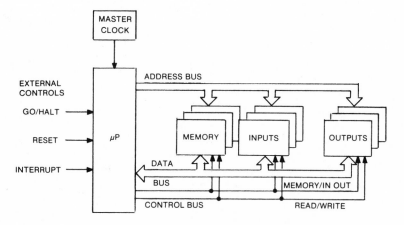

Figure 5.3 Basic microcomputer diagram.

The instruction is usually to do something with the data stored in the **data buffer**, which may have come from an input device or from memory. The instruction may be to perform an arithmetic operation, the result being stored temporarily in the **accumulator**. This is a temporary storage location, used as a kind of "running total" until the operation is completed, after which the result goes back into the memory address specified. Associated with the **arithmetic logic unit**, or **ALU** and accumulator is a set of **condition codes**, also called **status flags**. Each of these is a one-bit register that is set high or low (1 or 0) to indicate something about the result in the accumulator: its sign (+ or −), if it is all zeros, if there is a carry, an overflow, and so on. Five or six flags are the usual number.

Sometimes there is a frequently used subroutine in the program that requires several instructions, always in the same sequence. For added speed, these are stored in adjacent memory addresses and called a **stack**. Instead of each having to be accessed separately, they are addressed as if they were one memory location, and this address is stored in the **stack pointer**. The controller has to use only the single address to call for the entire stack.

The other registers shown in Figure 5.2 are called **general purposes registers**, to be used as required. There is an odd way of designating the registers in a controller. The A register is the accumulator, usually called accumulator rather than A register. Then there are the B, C, D, E, etc., registers, plus special-purpose ones such as an H register (high-order byte) and L register (low-order byte), or whatever takes the designer's fancy.

Figure 5.2 also indicates a number of external connections. These include a clock, power supply, data input/output, and so on. The clock is a generator for timing pulses with a repetition rate of one or more megahertz for synchronizing operations in the controller so it works properly. The power supply needs no explanation. The other lines into the control unit are all concerned with stopping or interrupting the program for the entry of external data, and restarting it afterwards.

The timing diagram in Figure 5.4 shows how the three buses

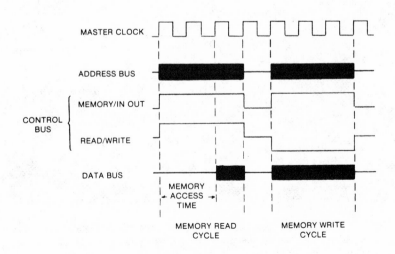

Figure 5.4 Timing diagram for microcomputer in Figure 5.3.

are synchronized. The microprocessor sets up the address and control buses on the leading edge of the next master clock pulse (the first in the figure). The internal memory location is transferred to the address bus, but nothing more can be done until the appropriate command is given on the control bus.

This is a command to select memory, given by a signal going high on the MEMORY/IN OUT line. At the same time, the READ/WRITE line also goes high, so the command is to read from memory. The microprocessor now looks for data on the data bus at the third clock pulse. The interval between the command and the appearance of the data is the memory access time, which is the time allowed for the reading cycle to be completed.

Two clock pulses later, a new address is specified on the address bus, while the control bus MEMORY/IN OUT line goes high and

(Courtesy Rockwell International, Inc.)

Figure 5.5 Magnified photograph of confetti-size chip (same size as similar chip in frontispiece) containing an entire microcomputer. By encoding different programs into the ROM (which is a comparatively low-cost procedure), the same microcomputer can be used for purposes as diverse as cash registers, electronic games, controls for machinery, and many more.

the READ/WRITE line goes low. This is a command to write data into memory at the new address given on the address bus, and another three pulses are allotted for this operation.

Input and output devices as indicated in Figure 5.3 are required to connect the microcomputer to the outside world. If it were not for these macro devices, the entire microcomputer could be housed in a wristwatch (see Figure 5.5). When voice input and output have been perfected, this could actually happen!

I/O Devices

The bulk memory*, keyboard, CRT, and printer have to be interfaced with the computer by circuits that translate the data from the format used in the computer to that used in the peripheral device, and vice versa. The following are descriptions of typical interfaces. However, each manufacturer uses interfaces that will be compatible with the microprocessor in his computer, so details may vary.

Diskette Controller

Although some microcomputers use tape cassettes, and many large computers use tape drives, the **magnetic disk** is widely employed, and its interface is therefore described here. Magnetic disks may be hard (expensive), or flexible (low-cost). The flexible disks are popularly called "floppy disks." They are available in different sizes: eight-inch diameter for business microcomputers, five and a quarter-inch diameter for standard personal computers, and three-inch for more compact personal computers. The eight-inch disk is usually termed a diskette, the smaller sizes "minifloppies." The following remarks apply mainly to diskettes, although they are much the same for any type; they are included here to help with understanding what the disk controller interface has to do.

Standard diskettes have 77 tracks, numbered from 0 to 76, starting at the outside edge. Generally, only one side of the diskette

---

*Not the internal memory.

is used (single-sided), but some use both sides (dual-sided). Data may be recorded in either single-density format (3408 bits per track), or more usually in double-density format (6816 bits per track), as magnetic patterns of 0s and 1s (as on a tape) along each track. These are structured in sectors. As a rule, there are 26 sectors per track, numbered from 0 to 25, with 256 **bytes*** per sector. Thus, the total capacity of the diskette is more than 500,000 bytes. A whole sector is always read or written at a time, and all data on the diskette is identified by sector and track number. The read-write head of the disk drive can access each track, by moving in or out radially, and each sector, as the diskette rotates under it. (The rate of rotation is 60 times a second.)

Diskettes can be hard-sectored or soft-sectored. In the first-named, each sector has an index hole punched in a special track to indicate its position. The location of this hole is read by a photo device as the hole passes in front of a LED. In the more common soft-sectored diskette there is only one index hole, marking the beginning of the first sector. The positions of the other sectors must then be computed from this hole's location, another job for the diskette controller.

Figure 5.6 shows both what goes into each sector, and the layout of the diskette controller IC. In the sector are an ID field and a data field. The ID field is to enable the controller to position the read-write head correctly over the desired sector. The various parts of the controller help the computer microprocessor write and read data into and out of the diskette.

The **direct memory access** (DMA) control logic provides for transferring data directly between the diskette and the computer's internal memory without going through the CPU (microprocessor). This is faster. The DMA controller takes over control of the address, data, and control buffers from the CPU while the operation is carried out.

The **interrupt control** is used by the diskette controller to notify the CPU that it wishes to interrupt the main program. The main program can then respond to the interrupt message when convenient. The **read/write control** processes the input from the

---

*See Glossary.

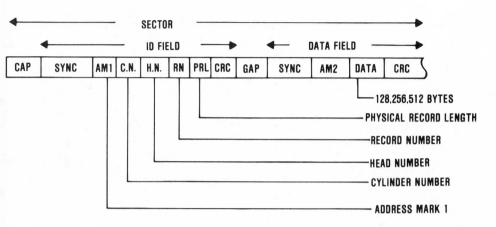

(a) SECTOR FORMAT

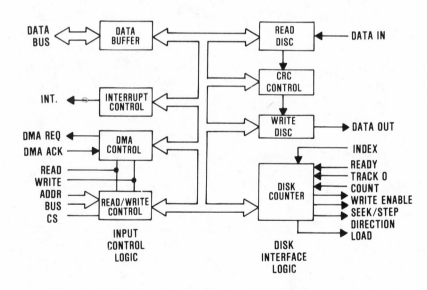

(b) DISKETTE CONTROLLER IC

After Encyclopedia of Integrated Circuits, *W.H. Buchsbaum, Prentice-Hall, 1981, page 321.*

Figure 5.6 (a) Sector Format (b) Diskette Controller IC

microprocessor when the latter signals to activate the diskette controller.

The **disk interface logic** includes the **read disk logic**. This handles the data read from the diskette. It routes the data from the different parts of the sector to the corresponding sections of the controller.

For instance, the contents of the **cycle redundancy check** (CRC) bits in the ID part of the sector, are compared with the corresponding bits in the CRC logic. If they are not the same, the CPU is notified that an error has been detected, and the CPU displays an appropriate error message (for example, CRC ERROR DURING DISK I/O OPERATION).

The **write disk logic** writes the data from the CPU on to the diskette. The **disk controller** or **disk counter** moves the read/write head from track to track, counting the tracks as it goes, and lowers ("loads") it on to the selected track. The index input is the signal generated each time an index hole in the diskette is sensed.

It may seem that the diskette controller has a lot of freedom, but in fact it is at all times under the control of the CPU (even in those cases where a separate microprocessor is provided), which in turn is controlled by the **disk operating system**. This is a program which is loaded into the computer's internal memory at the start of operations. It contains the instructions for formatting a new disk with tracks and sectors, copying one disk on to another, deleting material, accessing files, detecting errors, and also providing system routines for controlling the keyboard, video display, printer, and many more.

Keyboard Interface

The standard keyboard has movable keys that rest on light coil springs. When depressed they close contacts that result in binary signals being sent to the CPU. When a key is depressed, a **key-activate** signal is sent to the interface IC, as shown in Figure 5.7. This is detected by the **keyboard detector and debouncer** circuit, which waits for a brief moment to allow the circuit to stabilize, and then checks the keyboard-activate input again.

Meanwhile, the **scan counter** continually scans the keyboard

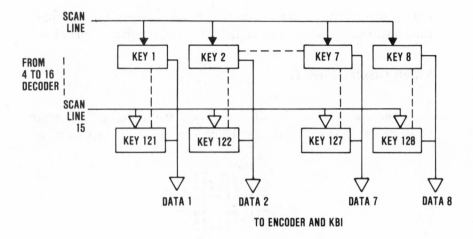

a) KEYBOARD OPERATION

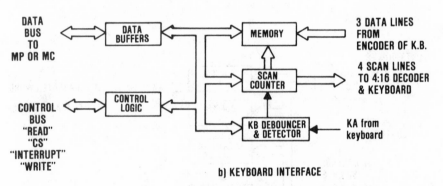

b) KEYBOARD INTERFACE

After Encyclopedia of Integrated Circuits, *W.H. Buchsbaum, Prentice-Hall,
1981, page 331.*

Figure 5.7   (a) Keyboard Operation (b) Keyboard Interface

with signals sent out over up to 16 scan lines. Each of these lines is
connected to eight keys. When a key is depressed, it generates a
3-bit word on the data line connected to it, while the scan line connected to it adds another 4 bits. In this way, a 7-bit word unique to
that key is stored in the interface *memory*. (This memory can actually store several words if necessary.) The **control logic** circuit sends an
interrupt signal to the CPU, and if the CPU is not otherwise en-
gaged it reads the word stored in the memory, canceling the inter-

rupt as soon as the memory is empty. However, some keyboards have their own microprocessor to free the CPU from keyboard operations.

Video Display Interface

The operation of the CRT is similar to that of a TV picture tube, except that each horizontal line consists of discrete dots, in-

(a) DISPLAYED CHARACTER IN DOT MATRIX FORM

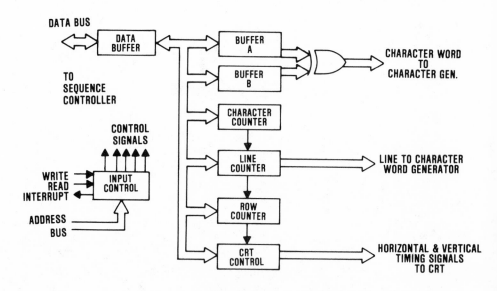

(b) CRT CONTROLLER INTERFACE

After Encyclopedia of Integrated Circuits, *W.H. Buchsbaum, Prentice-Hall, 1981, page 318.*

Figure 5.8   (a) Displayed Character (b) CRT Controller Interface

stead of varying brightness modulation. Typically, the screen can display 24 rows of 80 characters, the characters being formed from a seven by eight-dot matrix.

As shown in Figure 5.8, the CRT interface accepts data from the CPU via the data bus, and stores it in two **buffers**. Each buffer holds one row. One buffer transfers data to the CRT while the other receives the next row. The number of characters in each row is programmable, so they are counted by the **character-counter** circuit until this number is reached, and then the other buffer is activated and the next row is displayed. Characters are displayed as they are received.

The **line counter** counts the horizontal lines required by each row. These include the eight used by the matrix plus those between the rows. After counting the correct number it resets to zero. Meanwhile, the **row counter** is updated by one count for each set of horizontal lines. When it reaches the maximum number of rows (24) for the screen, it will start to "scroll." This means it will move the display up one line at a time as long as extra lines are added. For each extra line added at the foot of the screen, a line disappears from the top of the screen, so there can never be more than 24 on the screen at one time. (This part of the display can be brought back on the screen by scrolling in the opposite direction with one of the control keys provided on the keyboard.)

The **CRT control** circuit keeps the vertical and horizontal sweeps synchronized. This function is the same in principle as that in a TV set.

All the foregoing logic functions are under the control of the **input control** circuit, which interfaces with the CPU.

Printer Interface

Different types of printers require different types of interface, so the circuitry is in the printer rather than in the computer. The cable connecting the computer to the printer typically has 16 wires. Eight of these are for data, and two more for clock and reset signals to the printer. Five are for signals from the printer to the computer, and the last is for the common signal ground.

The data are sent to the printer in **ASCII** (American Standard

Code for Information Interchange) or TTY (teletype) code. The ASCII code consists of words of seven bits, not all of which are used or are applicable to the printer being used. (See Table 5.1.) The signals sent from the printer to the computer consist of *acknowledge* (receipt of data byte), *busy, paper empty, select* (device selected), and *fault* (ribbon cassette empty, printer cover removed, etc.).

CHARACTER	CODE	CHARACTER	CODE
A	65	2	50
B	66	3	51
C	67	4	52
D	68	5	53
E	69	6	54
F	70	7	55
G	71	8	56
H	72	9	57
I	73	+	43
J	74	—	45
K	75	/	47
L	76	*	42
M	77	↑	94
N	78	(	40
O	79	)	41
P	80	<	60
Q	81	>	62
R	82	=	61
S	83	?	63
T	84	$	36
U	85	"	34
V	86	'	44
W	87	.	46
X	88	;	59
Y	89	CARRIAGE	
Z	90	RETURN (CR)	13
0	48	LINE FEED (LF)	10
1	49	SPACE	32

Table 5.1 Partial Listing of ASCII

In Figure 5.9 the printer interface IC from a dot-matrix printer is shown. This printer operates by advancing selected rods that strike the paper to form characters as shown. The paper is marked by ink ribbon, heat, or other means. However, other printers can print a whole line at one stroke (line printers), while some perform like a standard electric typewriter, using a type ball or daisy wheel.

a) DOT MATRIX CHARACTER

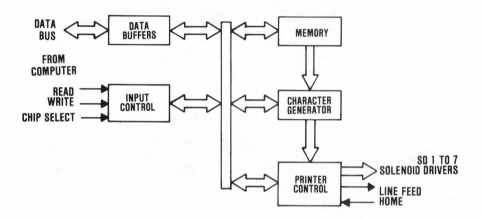

b) PRINTER INTERFACE IC

After Encyclopedia of Integrated Circuits, *W.H. Buchsbaum, Prentice-Hall, 1981, page 333.*

Figure 5.9   (a) Dot Matrix Character (b) Printer Interface IC

The line printer is very fast (10 lines a second), the dot-matrix moderately fast (20 to 60 lines per minute), and the daisy-wheel type comparatively slow (45 to 55 characters per second). The latter is used mostly for word processing, because of the typewriter-like quality of the print. It goes without saying that the different methods of printing require different character-generator and printer-control logic.

The computer uses the binary equivalent of these code numbers. For instance, decimal 65 is converted by the computer to the seven-bit binary number 1000001. In this condensed list, lower-case letters and many other characters and instructions have been omitted.

CHAPTER 6

Most Amplifier ICs Are Differential or Operational

As you have seen in the previous chapters, digital circuits are logic circuits that carry out binary arithmetic functions and other processes by using two voltage levels, true and false, to operate electronic switches called gates.

Gates, like other switches, have only two states: "on" and "off." They are not continuously variable. To handle signals that are the electrical analogs or equivalents of fluctuating quantities, such as temperature, intensity of illumination, music, and so on, we must use **linear amplifiers**.

Linear Amplifiers

A linear amplifier is an amplifier biased to place its operating point midway between cutoff and saturation, so that excursions of an input signal are confined to the linear portion of the forward voltage characteristic curve of the transistor, as shown in Figure 6.1. This results in an output signal that is a faithful replica of the input signal, apart from being amplified.

Direct-Coupled Amplifiers, Including Darlington-Coupled Amplifiers

A linear amplifier constructed with discrete components could have a circuit similar to that shown in Figure 6.2. To duplicate this

129

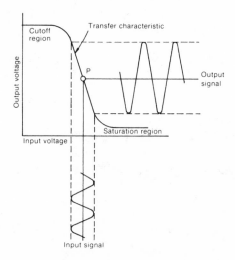

Figure 6.1   Linear amplification.

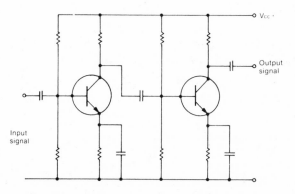

Figure 6.2   Two-stage discrete transistor amplifier.

in the form of an IC would be difficult and expensive because the high values of the resistors and capacitors would not lend themselves to monolithic fabrication. As you saw in earlier chapters, capacitors on chips are limited to a maximum of a few hundred picofarads, and the use of higher-value resistors is not economic. Consequently, circuits resembling those in Figure 6.3 are used, where direct coupling eliminates the need for capacitors, and resistors with low values are used. The total number of components in the circuits in Figure 6.3

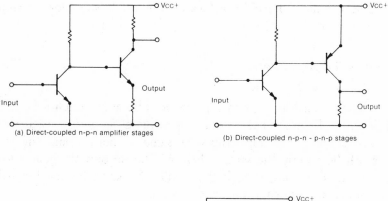

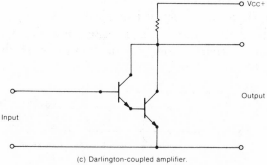

(c) Darlington-coupled amplifier.

Figure 6.3   Direct-coupled transistor amplifiers.

are five, four and three respectively, compared with 15 in Figure 6.2. The Darlington-coupled amplifier is often used because of its high amplification, which can be several thousand times. However, there are problems with these circuits, the worst of which is DC drift.

DC Drift

Because these direct-coupled circuits lack the resistor networks of Figure 6.2, which prevent the operating points of their transistors from changing with temperature variation, a small change in the operating point of the first stage results in an amplified change in its output voltage. With no coupling capacitor to isolate this from the input of the following stage, this voltage will appear on the base of the next transistor, to be amplified further by this and every succeeding stage. The final DC output level may be considerably altered as a result. To avoid this, nearly all linear ICs employ **differential amplifiers**.

Differential Amplifiers

The schematic diagram for a basic differential-amplifier circuit is shown in Figure 6.4. It consists of identical amplifiers, each with its own input. If you apply the same signal to both inputs, the outputs will be exactly the same. There will be no output signal across the output terminals, though there will, of course, be equal signals between each output terminal and ground.

However, if the input signal is applied to one input only, the current through that transistor will increase when the signal swings positive. As the current through both transistors is the current

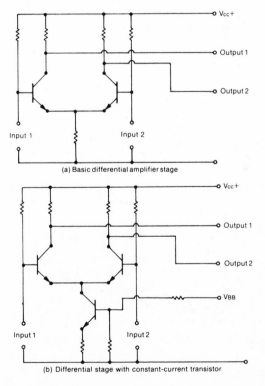

(a) Basic differential amplifier stage

(b) Differential stage with constant-current transistor

Figure 6.4   Differential-amplifier stages.

through their common emitter resistor, which tends to remain constant, the current through the other transistor decreases. The collector of the first transistor becomes less positive, but the second's collector becomes more positive. This gives an output signal across the two output terminals. (We can also use the signals between either output point and ground; they will be of opposite phase.)

The importance of this circuit lies in the fact that any signal that is common to both sides is balanced out. This includes DC drift, hum pickup and other unwanted interference, which affect both amplifiers equally.

To work well, it is necessary for the current through the emitter resistor to be as constant as possible. This can be done by using a resistor with a high value but, for the reasons given previously, this is not a good solution. A better way is to replace it with a transistor. This can be biased with a fixed voltage, V_{BB}, so that it is in effect a resistor with a fixed value, which can be as high as necessary.

Common-Mode Rejection

Signals common to both sides of the circuit are called **common-mode** signals, and the ability of the amplifier to balance them out is called **common-mode rejection**. The value of this is usually expressed as a ratio in decibels. If a differential amplifier is said to have a common-mode rejection ratio (CMRR) of 80 dB, it means that any common-mode signal will be attenuated by a factor of 10,000. Such a figure is quite common, as the two amplifiers and their load resistors are formed simultaneously on the chip, and therefore will match very well.

Another advantage of using a transistor constant-current circuit is that, in some applications, negative feedback can be used from a subsequent stage to vary the resistance of the transistor, which can increase the common-mode rejection further. We shall come back to the importance of feedback when we discuss operational amplifiers.

Most linear amplifiers are audio amplifiers. They have a wide range of applications. For instance, in a hearing aid, a differential amplifier with four stages driving a class B push-pull output can be fabricated on a chip not much over a millimeter square. Powered by a 1.5-volt battery, such an amplifier may have a voltage gain of 4000.

Operational Amplifiers

In the last chapter we discussed various ICs that are used in *digital* computers, that operate by the presence or absence of discrete signals. Another type of computer is called an *analog* computer because it deals with various physical conditions, such as velocity, acceleration, angular position and the like, that are continuously variable. The analog computer was particularly suitable for handling these functions, in ballistic missiles.

The amplifier used to carry out these operations was called an **operational amplifier**. This was a linear amplifier whose output current was proportional to its input voltage. By causing the output current to flow through a suitable network, it could be made to bear any required relationship to the input. Today, this versatile device is used for a great number of other purposes, as we shall see.

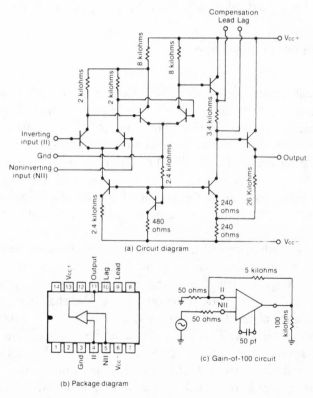

Figure 6.5 General-purpose operational amplifier.

Great precision and accuracy are demanded of the operational amplifier. As we have just mentioned, one way to achieve these qualities in monolithic circuits is by use of differential amplifiers. We also mentioned that the differential amplifier's stability could be further enhanced by providing negative feedback. This produces an amplifier with a precise gain characteristic depending only upon the feedback used.

A typical operational amplifier circuit is shown in Figure 6.5. This is a medium-gain amplifier, with two differential-amplifier stages, with constant-current transistors, but only one output is used. This goes to a driver stage, and then to an emitter-follower output stage.

Operational amplifiers are designed to be used with *external* negative feedback, as shown in Figure 6.6. The type of feedback net-

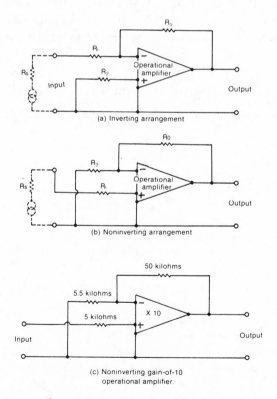

(a) Inverting arrangement

(b) Noninverting arrangement

(c) Noninverting gain-of-10
operational amplifier.

Figure 6.6 Operational amplifier circuits.

work will vary somewhat with the purpose for which the amplifier is required. In fact, the same amplifier can be used for a variety of purposes according to the external circuitry provided. This circuitry is quite simple.

Without it, the amplifier would have a very high *open-loop* gain, but to achieve maximum stability, part of the output signal is fed back to the input. At the input, the output signal is of opposite phase to the input signal, consequently some of the input signal is canceled out, resulting in a much lower *closed-loop* gain. This sacrifice of gain is necessary to obtain the superior operating characteristics required.

The input signal can be applied to either input. One input is called the *inverting input* (II). The output signal is of opposite phase to this input signal. The other input is called the *non-inverting input* (NII), because the output signal is in phase with the input signal at this terminal. It hardly needs saying that the feedback signal nearly always goes to the inverting input (otherwise it would be positive feedback). You'll also recall that whichever input the input signal is applied to, an equal signal of opposite phase appears on the other side because of the common constant-current emitter circuit.

In Figure 6.6 the feedback elements are R_o and R_i. The ratio $\dfrac{R_o}{R_i}$ gives the closed-loop gain, $\dfrac{E_o}{-E_i}$ (E_i is negative because the inverting

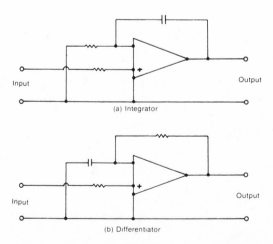

(a) Integrator

(b) Differentiator

Figure 6.7   The use of operational amplifiers for analog computing functions.

input and the output are opposite phase). For instance, if R_o is 5 kΩ and R_i 50 Ω, the closed-loop gain is $\frac{5000}{50}$ = 100. This value is practically independent of the open-loop gain, making it possible to obtain a precise amplification ratio by choice of suitable resistances for R_o and R_i.

If a capacitor is used as a feedback element, as shown in Figure 6.7, the gain is given by the ratio of the impedance of the capacitor to R_i,

$$\frac{E_o}{-E_i} = \frac{XC_o}{R_i} \tag{1}$$

However,

$$XC_o = \frac{1}{2\pi fC_o} \tag{2}$$

Rearranging (1) to find E_o gives:

$$E_o = \frac{-E_i XC_o}{R_i} \tag{3}$$

and substituting (2) into (3) results in:

$$E_o = \frac{-E_i}{R_i(j2\pi fC_o)} \tag{4}$$

(note operator j has to be used to denote that XC_o is reactive).
This expression can be integrated to give:

$$E_o = -\frac{1}{R_iC_o}\int E_i \frac{d}{dt} \tag{5}$$

This tells us that the output is an integral of the input, so with this type of feedback network, the operational amplifier becomes an *integrator*.

Since differentiation is inverse integration, we can convert the integrator to a *differentiator* by interchanging the capacitor C_o and the resistor R_i, which now become C_i and R_o.

The equation for the gain now becomes:

$$E_o = -R_o C_i \frac{dE_i}{dt} \tag{6}$$

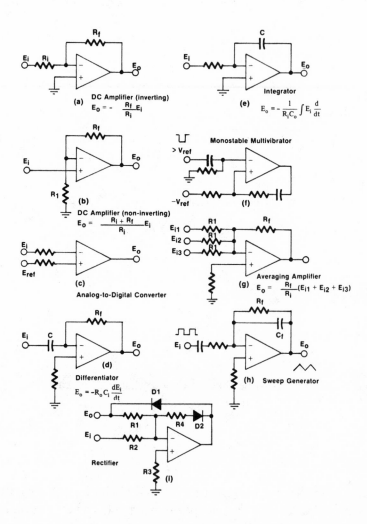

Figure 6.8 Various op-amp circuits.

Integrators and differentiators are used in the computation of velocity, acceleration, work done by a force, lengths of curves, and so on. Some other operational amplifier circuits are shown in Figure 6.8.

Wideband Video Amplifiers

The amplifiers discussed so far have been low-frequency or audio amplifiers, but other amplifiers operating up to microwave frequencies are also in use. Among these the most important is the wideband video amplifier.

This amplifier may be single-ended or differential but the differential is more efficient for the reasons given previously for lower frequency amplifiers. Although the differential amplifier has additional components, these add little to the cost of the IC. A typical example is illustrated in Figure 6.9.

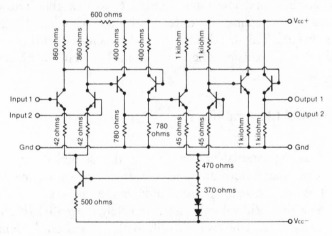

Figure 6.9   Circuit diagram for wideband video amplifier.

This amplifier consists of four stages, an input differential stage with common-mode feedback, coupled by emitter followers to another differential stage and emitter-follower output circuits. It is suitable for radar pulse amplification, having a flat frequency response from DC to 40 MHz. Similar amplifiers for television would only have to go to 5 MHz (black-and-white) or 10 MHz (color).

The upper frequency limit is set by the transistor and load resistor characteristics. As you know, the reactance of the distributed capacitance in the circuit decreases as the frequency rises, and as this is in parallel with each load resistor it reduces its value, and so cuts down the voltage gain of the transistor. The distributed capacitance includes the transistor collector capacitance, so that the type of transistor makes a considerable difference to the total circuit capacitance. In the amplifier in Figure 6.9 the total distributed capacitance is about 5 pF. At 40 MHz, the capacitative reactance of this capacitance will be:

$$X_c = \frac{1}{2\pi fC} = \frac{1}{2 \times \pi \times 40 \times 10^6 \times 5 \times 10^{-12}} \approx 800 \ \Omega$$

Consequently, the gain of a transistor with a load resistor of this value will be reduced by 30 percent (3 dB down) at this frequency.

The emitter-follower circuits are used because they have low input capacitance and low input resistance, so that they provide the best means of coupling the differential amplifier without degrading the bandwidth. This bandwidth could be increased further by addition of external feedback (the feedback in the circuit in Figure 6.9 is internal).

Narrowband I.F. Amplifiers

By adding external tuning circuits, this amplifier can be made into a narrow-bandwidth amplifier. External circuits are required because it is not possible to form inductances on a silicon chip (at least, not with useful values), so a completely integrated IF amplifier is not practical yet. However, the differential stages can be fabricated monolithically with external connections for IF transformers, crystal or ceramic filters, producing an IF amplifier with very superior performance.

Differential Comparators

Among other applications of linear amplifiers is the **differential comparator**. This amplifier is used to compare two voltages or signals, and to amplify the difference between them. One input is usually a reference, and as long as the other input is of the same value there will be no output. If, however, a voltage difference develops between the two inputs, this difference will be amplified, and appear at the output as an error signal.

Industrial Controls

The differential comparator is very important to the control of industrial equipment. It is no secret that power-line voltage, for instance, fluctuates considerably during the daily cycle. This makes lights vary in brightness, heaters in temperature, and machines in speed. In many processes no variation can be tolerated, so a means of maintaining a constant level must be provided.

An illumination control system (Figure 6.10) uses a photoconductive device in one arm of a resistance bridge. This bridge is adjusted so that it is balanced when the illumination is at the correct intensity. If the intensity changes, the resistance of the photocell changes, unbalancing the bridge. The two sides of the bridge are connected to the two inputs of the comparator.

The output of the comparator is connected to the input of the phase control unit. This unit produces a sawtooth signal that is synchronized with the power-line voltage so that each ramp coincides with one half-cycle of the sine wave. A feedback circuit applies this sawtooth to the reference input of the phase control unit's internal comparator.

In the absence of the sawtooth, this input would be at the potential set by a divider chain between $+V_{cc}$ and $-V_{cc}$. The beginning of the sawtooth ramp drives this input down below this level, and then it rises linearly to the higher potential at the end of the ramp, which is above this level. At some point, the sawtooth voltage equals that of the DC input voltage from the bridge comparator.

The pulse generator is designed to generate a pulse when this condition occurs, and this pulse triggers the triac into conduction, so that current flows through the lamp or lamps whose illumination intensity is being controlled. If the brightness of the lamp decreases, the photocell voltage decreases, so that the bridge balances earlier and the pulse trigger is generated earlier. This turns the triac on sooner, so that more current flows on each half-cycle. Conversely, if

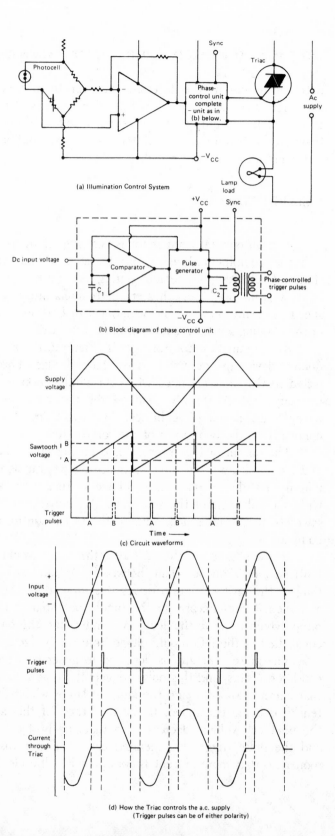

(a) Illumination Control System

(b) Block diagram of phase control unit

(c) Circuit waveforms

Figure 6.10
Typical industrial
control centers.

(d) How the Triac controls the a.c. supply
(Trigger pulses can be of either polarity)

the brightness increases above the set point, the photocell voltage increases, and the bridge is balanced later. This causes the triac to be triggered later, so that current flows for a shorter time on each half-cycle.

This method of control may be used to regulate the power in many other applications by substituting a suitable transducer for the photocell in Figure 6.10. For temperature control, that would be a thermocouple, for speed control, a DC tachometer. However, for speed control, there is also a digital system in use, in which a train of pulses generated by the machine is compared with a similar train from a crystal oscillator. Any phase difference between the two pulse trains causes an advance or delay in the pulse triggers controlling the triac, until the two signals are again synchronized.

The foregoing discussion by no means covers all the uses of linear ICs in industrial control. Most of the applications started as discrete-component assemblies, but the introduction of ICs resulted in higher reliability and better performance at lower cost. This has lead to the extension of electronic controls into the areas of consumer appliances and the automobile.

Power Amplifiers

Power amplifier ICs are available with output power up to 15 watts. These are generally Class B, and may have two channels for stereo. For many receivers, the IC frequently contains the entire audio section (for instance, the GE ICI80 used in some TV receivers).

The heat sink mounting and heat dissipation are very critical in power amplifier ICs, which often have automatic shutdown for excessive temperature, current or voltage.

For power output circuits of higher wattage, discrete power transistors must be used.

CHAPTER 7

But Today's Linear ICs Include More Than Amplifiers

Video and Sound Detectors

While video and sound detectors may be obtained as single-purpose ICs, they are more often found as subcircuits in system ICs that contain IF and other amplifiers as well. An example of this is shown in Figure 7.1, which shows the detector portion of circuitry on a PIX-IF-subsystem chip.

This consists of transistor Q2 and its biasing network Q1, Q3, and R2. Transistors are better for AM detection than diodes, because they are less noisy and less susceptible to broad-band IF interference. Q1 is biased to the same potential as Q2 because their bases are tied together through the resistance of the low-pass filter that consists of R1 and C1. R2 and C2 form a conventional peak detector in which the time constants are selected for optimum detector efficiency and desired video bandwidth.

A multimode detector IC is one that contains detection circuits for AM, FM, SSB, and CW signals. As shown in Figure 7.2, such a chip may contain three amplifiers in addition to the detector circuit. Amplifier 1 is an independent audio amplifier that is not connected to the other circuitry. It is provided for use as an audio amplifier if required. Amplifier 2 is a limiter for use when FM signals are to be detected. Amplifier 3 is a gain-controlled output amplifier that can be used with a filter connected to its input, if desired.

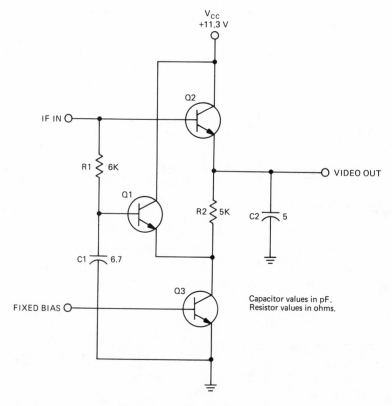

Figure 7.1 Video Detector Circuit.

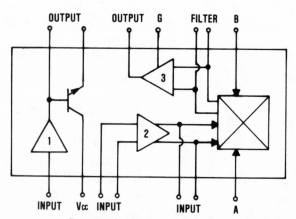

From Encyclopedia of Integrated Circuits, *W.H. Buchsbaum, Prentice-Hall, 1981, page 33.*

Figure 7.2 Multimode Detector IC.

Phase-Locked Loop (PLL)

The phase-locked loop is a very versatile circuit that can be used for detection, as well as numerous other purposes. As shown in Figure 7.3, it contains a voltage-controlled oscillator (VCO). In the absence of an input signal, the VCO runs at its own frequency, which is applied to the phase comparator. When an external signal is also applied to the phase comparator the latter compares its frequency to that of the VCO and generates an error signal if they are different. This error signal is fed back via a filter to the VCO, which shifts its frequency to agree with that of the incoming external signal, and locks on to it. When used as an FM detector, the variations in the error signal become the recovered audio signal.

PLLs may also be used for FM stereo decoding, motor speed control, signal conditioning, AM and FSK detection, and many other applications.

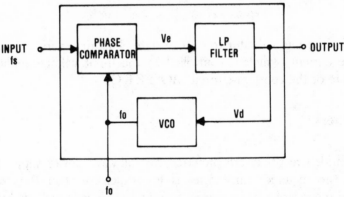

From Encyclopedia of Integrated Circuits, *W.H. Buchsbaum, Prentice-Hall, 1981, page 45.*

Figure 7.3   Phase-Locked Loop.

Voltage-Controlled Oscillator (VCO)

A VCO on a chip (Figure 7.4) uses a multivibrator (gates 1 and 2), whose frequency is determined by the time taken for a constant current to charge an external capacitor C1 to the level neces-

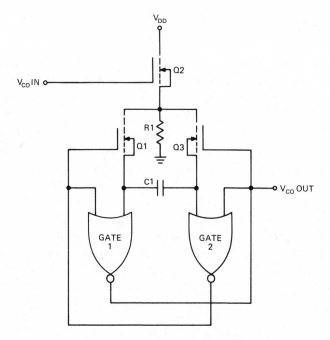

Figure 7.4   Partial Schematic of VCO.

sary for the multivibrator to change state. This time varies according to the current, which is controlled by the input voltage applied to the gate of the constant-current MOSFET Q2.

Rectifiers

Bridge rectifiers are preferred for full-wave power supplies because the required transformer is less expensive than the center-tapped secondary type used with two diodes. Since four-diode bridges are available as ICs, the only question to be considered is the power-handling capability of the IC. Be sure to check that the bridge is capable of handling the maximum amperage and peak inverse voltage (PIV) called for.

Voltage Regulators

A voltage regulator in an IC operates in the same way as one

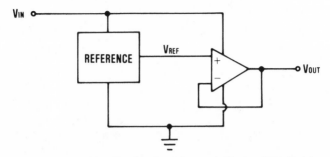

From Encyclopedia of Integrated Circuits, *W.H. Buchsbaum, Prentice-Hall, 1981, page 52.*

Figure 7.5 Basic Voltage Regulator IC. Output frequently drives a power transistor, or Darlington pair, for higher current capability.

built of discrete components. As shown in Figure 7.5, the basic regulator consists of an operational amplifier and a reference. The latter may be a zener diode, or a more complex transistor circuit.

Since the circuit illustrated has a limited power handling-capability, it may also be used to control an external series power transistor, Darlington pair, or some other configuration with higher current capability. The IC may also require a heat sink, and be provided with thermal shutdown, and overvoltage and overload protection.

Analog-to-Digital Converter

The A-D converter changes analog signals to digital. A signal from a transducer, such as a thermocouple, must be "digitized" before the data can be processed by a computer. Similarly, a voltage must be changed to digital form to drive the readout of a digital voltmeter. This requires that the output of the converter consists of a number of bits, according to the requirement of the computer or meter. Music is digitized also, to eliminate noise and to permit the addition of other desirable features in its reproduction. For this, the principle of **pulse code modulation** (PCM) is used.

Figure 7.6 is a block diagram of an IC used in 3½-digit multimeters and other devices where the readout consists of three figures with or without a leading 1 as well. The analog signal is fed to the analog subsystem, where it is compared with the reference

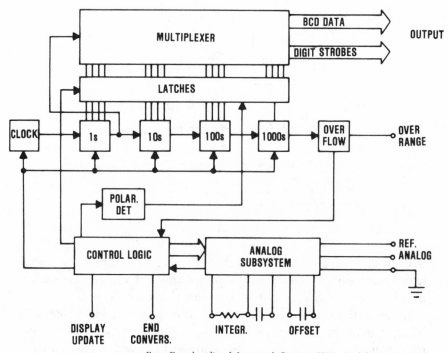

From Encyclopedia of Integrated Circuits, W.H. Buchsbaum, Prentice-Hall, 1981, page 292.

Figure 7.6 Analog-to-Digital Converter for DMM.

voltage, using a dual-ramp converter. In this, as shown in Figure 7.7, the analog voltage is applied to the input amplifier, which produces an output current I_X. This current is routed by the ramp control electronic switch to the integrator. The integrator output is a positive-going ramp corresponding to the amplitude of the analog voltage, which is allowed to run for a precise period of time (T1). At the end of this time, the switch connects the integrator to the reference current (I_R). The resulting ramp is negative-going, and it reaches zero in a length of time (T2) determined by its initial amplitude. The comparator puts out a gating signal equal to T1 + T2. While this gating signal exists, pulses generated in the digital logic section are allowed to appear at the digital output, but are cut off when the down-going ramp hits zero. During the gating period the clock drives the four decade counters, and their outputs are processed by the latches and multiplexer to a form suitable for application to a BCD-

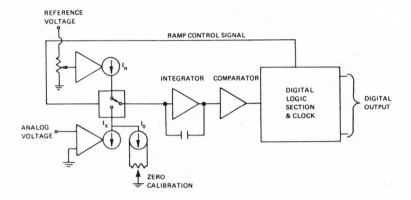

a) BLOCK DIAGRAM — DUAL RAMP A/D CONVERTER

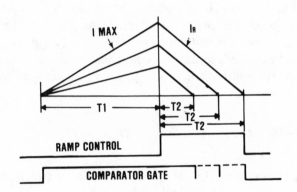

b) DUAL RAMP WAVEFORMS

From Encyclopedia of Integrated Circuits, *W.H. Buchsbaum, Prentice-Hall, 1981, page 287.*

Figure 7.7   Dual-Ramp A-D Converter.

to-7-segment converter (in some cases, the latter is available on the same chip).

Conversion of audio to PCM may be performed in a similar circuit. The fluctuating amplitude of the audio signal is sampled by dual ramps in the same way, resulting in bursts of pulses. The num-

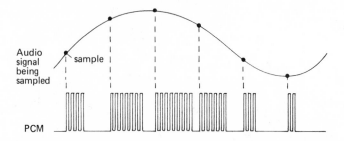

Figure 7.8 Conversion of audio to PCM results in bursts of pulses. The number of pulses in each burst corresponds to the amplitude of the sample.

ber of pulses in each burst corresponds with the amplitude of the sample (see Figure 7.8). The number of bursts per cycle depends on the frequency. Since the pulse rate is much greater than the audio frequency, each of its fluctuations is divided into a large number of pulses. (Instead of PCM, some A-D converters produce delta modulation in which the width of successive pulses varies with the slope of the analog waveform.)

Digital-to-Analog Converter

For the opposite process of D-A conversion, a circuit such as that in Figure 7.9 is used. Eight bits in parallel are applied to the switch drivers, which actuate the electronic switches as shown. The bias circuit ensures that a regulated voltage is applied to the reference amplifier and the resistors in the resistor network. The values of these resistors are such that a current corresponding to the binary value of each bit in the input, is connected to the output to give a total analog current that is the sum of the eight binary currents.

In decoding a PCM signal, the resistor network is replaced by an integrator that sums each burst of pulses to give an output corresponding to the amplitude of the original audio signal sample. Although these samples are pulses themselves, they are so numerous that they blend together to form an audio signal that is indistinguishable from the original.

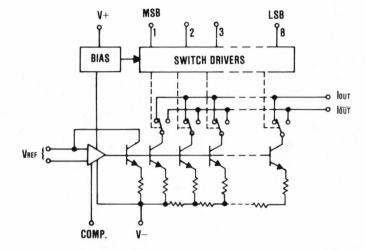

From Encyclopedia of Integrated Circuits, *W. H. Buchsbaum, Prentice-Hall, 1981, page 300.*

Figure 7.9   Digital-to-Analog Converter.

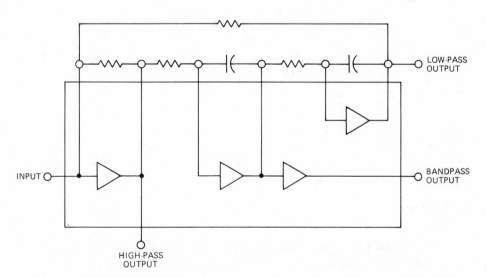

Figure 7.10   Active Filter IC. Values for the external resistors and capacitors must be chosen according to type of filter, frequency, etc.

Active Filter

Active filters use operational amplifiers with various feedback arrangements to replace the coils and capacitors of traditional passive filters. An IC for use as an active filter is shown in Figure 7.10. It can be a low-pass or high-pass filter, or a band-pass filter. Its characteristics are determined by the external resistors and capacitors connected to it. Other types of filters are also available.

Feedback Volume-Control Circuit

In television sets using integrated circuits, the volume control is frequently located remotely from the audio section. The long leads that must be used often pick up hum and noise at low volume levels. This requires the use of shielded leads unless a **feedback volume control** is used. As shown in Figure 7.11, this type of control is obtained by varying the *gain* of the audio amplifier, instead of varying the amplitude of the input signal, as is done in a conventional audio amplifier that runs at maximum gain all the time.

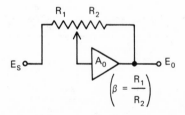

Figure 7.11   Feedback Volume Control.

When the wiper arm of the potentiometer is close to the input end of the resistive element, the gain of the amplifier is essentially the open-loop gain A_o. As the wiper arm is moved toward the output end of the resistive element, the gain is progressively reduced to closed-loop values (A_c) given by:

$$A_c = A_o /(1 + \beta A_o)$$

where β is the feedback factor (R_1/R_2).

Integrated-Circuit Diode Array

A typical IC diode array, as shown in Figure 7.12, consists of a diode-quad arrangement similar to a bridge rectifier circuit plus one

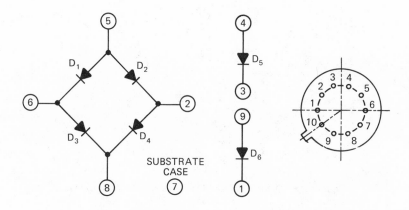

Figure 7.12   IC Diode Array.

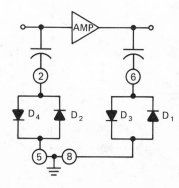

Figure 7.13   Diode Array Used in Limiting Circuit.

or more individual diodes. Its applications include gating, mixing, modulation, and detection. Because all the diodes are fabricated simultaneously on the same chip, they have nearly identical character-

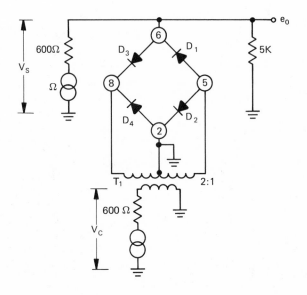

Figure 7.14   Diode Array Used in Balanced Modulator Circuit.

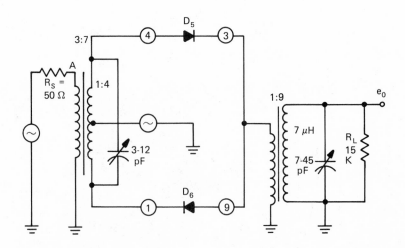

Figure 7.15   Diode Array Used in Balanced Mixer Circuit.

istics; and because of their close proximity, temperature variations affect them all equally. This makes them particularly useful in circuits requiring a balanced diode bridge or identical diodes.

Each diode is formed from a transistor by shorting together its collector and base for the anode and by using its emitter as the cathode, as you saw in an earlier chapter.

Figures 7.13 through 7.16 show various circuit arrangements using an IC diode array.

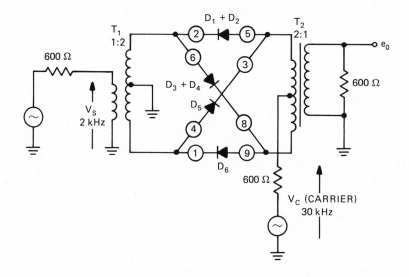

Figure 7.16   Diode Array Used in Ring Modulator Circuit.

CHAPTER 8

Microwaves
of the Future

One of the toughest items to microminiaturize has been radar. Some of its components hitherto have been quite massive. Even inside the radome of a fighter aircraft would be a steerable dish antenna with servo motors, synchros, pivoting waveguides, and other devices such as the transmitter's magnetron. Some of these also generated a great deal of heat.

Conventional Radar System

The principal assemblies in a conventional airborne radar system are shown in Figure 8.1. The synchronizer or timer originates synchronizing signals that trigger the modulator and the indicating device (the radar screen). The modulator emits a pulse that causes the magnetron to oscillate at the radar frequency for the duration of the pulse, and these bursts of energy travel along waveguides to the antenna. The duplexer prevents them from entering the receiver, but between pulses it connects the receiver to the antenna.

The antenna beams the pulses in the direction toward which it is pointed, swinging to and fro and up and down to scan its search area. When an echo comes back from a target, it is processed by the receiver and displayed on the screen. The screen is calibrated to show the target's range, bearing, and altitude, the second two items being reported by the antenna servos, which signal the attitude of the antenna at the time of the transmission of the pulse. These are

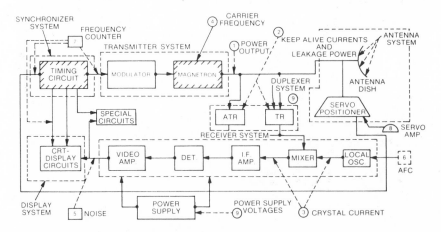

Figure 8.1 Conventional radar system (showing measurement points).

usually integrated automatically with the aircraft's heading and altitude to give a precise position for the target.

Most of the circuits employed in this radar are linear, since the signals being processed vary according to the quantity they represent. Digital circuits will be found only in the system's computer (assuming it has one). In microelectronics, however, many of these analog functions have been digitized, although not all as yet. Those involving the magnetron, the antenna and its mechanical equipment and electromechanical devices, the local oscillator and mixer of the receiver, the duplexer and waveguides, cannot be performed by logic circuits—at least, not so far.

Microelectronic Radar System

In a microelectronic radar some other method is needed to replace these items, which are anything but microscopic. Unfortunately, early attempts using silicon substrates were not too successful because manufacturing processes degraded the resistivity of the silicon and caused excessive dielectric losses. Ceramic substrates worked fine but were only a stopgap measure. They conducted heat well and were satisfactory insulators, but what the researchers were looking for was a single-crystal substrate on which active devices could be grown heteroepitaxially (see Chapter 1).

Sapphire is such a substance, but for maximum heat dissipation the choice soon shifted to **beryllium oxide**. Producing this in single-crystal form was a great breakthrough, for heat conductivity of beryllium oxide is approximately the same as that of brass.

Transmission Lines

Transmission lines are formed on the substrate by coating the underside with a layer of deposited metal as a ground plane, and depositing lines of metal called **microstrip** on the upper side to guide the microwave energy. Although vastly smaller than conventional waveguides or coaxial cable, these transmission lines have the same electrical properties, of which the most important is the characteristic impedance (Z_o). This is determined mainly by the ratio $\frac{b}{h}$ (see Figure 8.2) and the dielectric constant, although conductor thickness is a factor to some extent.

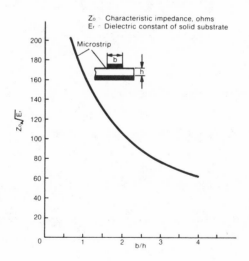

Figure 8.2 Microstrip structure and characteristic impedance.

As long as the ground plane is more than three times the width of the microstrip, the latter is regarded as extending to infinity. A Z_o of 50 ohms is given by a microstrip line where $\frac{b}{h} = 1.8$ and the dielectric constant is 5; if $\frac{b}{h} = 1$, with the same dielectric constant, Z_o is 75 ohms.

Interesting though the microstrip line is, it has a formidable competitor, the **microacoustic line**. This is made by depositing a strip of gallium arsenide on a beryllium substrate. Beryllium oxide, besides being an excellent insulator and heat conductor, it also piezo-electric—that is, it has the electromechanical properties of the quartz crystal. A microwave signal applied to this line is converted to an acoustic signal that travels along it piezoelectrically and is then re-converted to a microwave signal at the other end. The transducer at each end is a metallic pattern with fingers that are "interdigitated" as shown in Figure 8.3. Any potential difference between the fingers produces a mechanical strain that is transmitted along the line as an acoustic wave. At the other end, the mechanical vibration of the crystal reproduces the original signal across the fingers of the output transducer.

Figure 8.3 Microacoustic line.

It seems a little unnecessarily complicated to go to the trouble of converting a microwave signal to an acoustic one, and then back again, when it could have been transmitted directly with a suitable microstrip line. The reason for doing this is that an acoustic signal has much smaller dimensions, making it much more suitable for mi-croelectronic circuits. A one-gigahertz microwave signal, traveling at the speed of light, has a wavelength of 30 centimeters, as given by the equation:

$$\lambda = \frac{c}{f} = \frac{3 \times 10^8}{10^9} = 3 \times 10^{-1}\, m = 30\ cm$$

(λ = wavelength, c = speed of light in meters per second, f = frequency in hertz.)

An acoustic wave at the same frequency, but traveling at the slower speed of sound (v), has a wavelength given by

$$\lambda = \frac{v}{f} = \frac{3 \times 10^3}{10^9} = 3 \times 10^{-6} = 3 \ \mu m \ (microns)$$

This wavelength is therefore 100,000 times smaller than the microwave, which is an extremely valuable property for today's needs. A typical missile radar, for example, may be required to process in a few minutes hundreds of thousands of target echoes over a bandwidth of hundreds of megahertz. At microwave frequencies, a delay line to store these signals might have to be a kilometer long, but at acoustic wavelengths it would need to be only one centimeter.

Varactors

Microwave power may be generated by transistors, but these have rather low upper frequency limits in the megahertz range, which makes them unsuitable for radar. Another solid-state device that may be used is the **varactor** (variable reactor), which is a silicon junction diode doped to give large changes of capacitance between the forward-bias and reverse-bias condition. When a signal with a much lower frequency is applied to the varactor, the resulting capacitance changes cause a great many harmonics to appear in the output, from which the desired microwave frequency can be selected by tuning.

IMPATTs

The **impact avalanche transit time** (IMPATT) diode is also a silicon p-n junction diode. It is reverse-biased to its avalanche threshold. When an AC signal is superimposed on the biasing voltage, its negative-going and positive-going excursions will drive the diode into and out of the avalanche condition. When in the avalanche condition, a sudden breakdown comes about by the cumulative multiplication of hole-electron pairs due to field-induced impact ionization. At lower frequencies, all that happens is that the current flows on the negative-going half-cycles with the current and voltage in phase. As the frequency is raised, a point is reached where the electrons are still in transit when the positive-going half-cycle takes over. In other words, the current and voltage are now of opposite phase. Since this

condition offers negative resistance to an AC signal, sustained microwave oscillation breaks out from spontaneous fluctuations in carrier
movement during each breakdown induced by the applied signal.

Both IMPATT and varactor diodes are capable of a pulsed output of up to 10 watts at 10 gigahertz, and may be mounted on
stripline as shown in Figure 8.4. These devices are true p-n junction
diodes, and therefore have both majority and minority carriers that
can produce stored-carrier effects. These limit the upper frequency,
as in transistors. To avoid this, another type of diode with a metal-to-
semiconductor junction was developed for microwave operations.
This device is called the **Schottky barrier diode**, or "hot-carrier" diode. It is a majority-carrier device that operates like a varactor.

Figure 8.4   IMPATT or varactor diode mounted on stripline.

Gunn-Effect Diodes

However, it was soon replaced by the Gunn-effect diode, which
is not really a diode at all as it does not have any kind of junction,
but is a single crystal of gallium arsenide sandwiched between two
ohmic contacts, usually microstrip, as in Figure 8.5 (a). Gallium arsenide and some other crystalline compounds of elements from Group
III and Group V, or Group II and Group VI of the Periodic Table
of the Elements have a second conduction band with a higher energy level. At the proper threshold voltage, electrons from the first
conduction band are transferred into the second. For GaAs, the
threshold voltage is approximately 3000 volts per centimeter.

In the higher energy band, the effective mass of an electron is
considerably greater, so its mobility is reduced. The normal conduc-
tion-band mobility of GaAs electrons at room temperature is 9000
$cm^2/Vsec$, and their effective mass is 0.07 m_o (m_o is the mass of the
electron at rest). In the higher energy band, the effective mass increases to 0.4 m_o, and the mobility decreases to 150 $cm^2/Vsec$. As
the electrons drift through the crystal lattice, those with the lower
mobility bunch together to form a **field domain**.

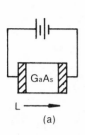

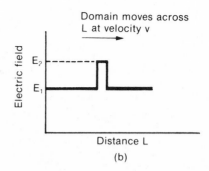

Figure 8.5 Gunn-effect diode: (a) structure and biasing; (b) formation of high-energy field domain.

This is illustrated in Figure 8.5 (b). Assuming a diode length L of 50 microns, the threshold voltage will be about

$$\frac{3000 \times 50 \times 10^{-6}}{10^{-2}} = 15 \text{ V}$$

As long as the potential across the diode is below this value, the electric field E_1 is distributed uniformly along the length of the diode. When the potential across the diode is raised to the critical value of 15 volts, electrons begin to transfer from the low-energy conduction band to the high-energy band. Since $v = \mu E$, where v is the drift velocity of the electrons (constant at about 10^7 centimeters per second), and μ is the mobility, a decrease in mobility requires a proportionate increase in E, so that the electric field potential increases with each electron transferred, which in turn increases the number transferring, resulting in a sudden and exponential increase in the electron population of the high-energy band. These electrons form a "shock wave" (high-energy field domain) that drifts from the cathode to the anode, where it disappears. Then, for a moment, the electric field in the diode reverts to E_1, the original state. But as the conditions are now the same as before, a new high-energy field domain is instantly formed, and the cycle is repeated.

Because of the conversion of so many high-mobility electrons from the low-energy band into low-mobility electrons in the high-energy band, the current through the diode decreases sharply with the formation of each high-energy field domain, and this diminished current persists until the domain disappears when, for a moment, the

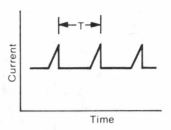

Figure 8.6   Gunn-effect diode current waveform.

original condition is restored. The current waveform, therefore, has a profile resembling that in Figure 8.6.

The time taken by each current pulse to travel from the cathode to the anode depends upon the distance L, which in Figure 8.5 (b) was arbitrarily chosen to be 50 microns. Since the velocity v is 10^7 centimeters (10^5 meters) per second, the time is given by:

$$T = \frac{L}{v} = \frac{50 \times 10^{-6}}{10^5} = 50 \times 10^{-11} = 5 \times 10^{-10} \text{ second.}$$

The frequency of oscillation is the reciprocal of this, of course,

$$\text{or } \frac{1}{5} \times 10^{10} = 2 \text{ gigahertz.}$$

To increase the frequency it is necessary to reduce the transit time, which means shortening the length of the diode. But as this is only 50 microns, there is not much room for improvement. A very thin diode can be obtained by epitaxial deposition, of course, and with beryllium oxide as a substrate the removal of heat is made much easier; there are practical limits to what can be done, however, as you can see.

LSA Diodes

Fortunately, by changing the doping it is possible to broaden the high-energy field domain so that it is as wide as the length of the diode, which is the same as if the diode were as narrow as the domain. Frequency is now determined by the rate of build-up and collapse of the domain alone, as transit time is nil. This mode of operation is termed **limited space-charge accumulation** (LSA). LSA diodes can be made much larger than Gunn-effect diodes, yet operate at the same or higher frequencies while delivering much higher power.

Phased-Array Antennas

By this time, you may be thinking it is all very interesting, but non-solid-state radar transmitters develop kilowatts, even megawatts of power, beside which the output of a microelectronic transmitter at present seems too puny for any but very low-power applications, and you are right. If coupled to a conventional antenna, the output power would be insignificant. A different approach is required.

The answer lies in the adaptation of the phased-array antenna to microelectronic use. This antenna consists of an array of dipoles, in which the signal feeding each is varied in such a way that antenna beams are formed in space and scanned very rapidly in azimuth and elevation. In the solid-state version, each dipole is mounted on the end edge of a substrate, on whose surface is deposited an entire radar transceiver for that dipole alone (see Figure 8.7). As many as 1600 of

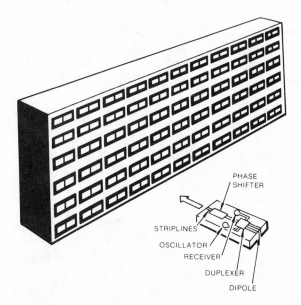

Figure 8.7 Principle of solid-state phased-array radar.

these elements form the phased array and may be assembled together in a flat planar configuration or installed conformally, which means arranged to follow the shape of some structure, such as an airplane wing.

This system is scanned in one of two ways. In the first, each element has a slightly different frequency; in the second, they all have the same frequency, but each has a small phase shift with respect to the others. Both versions have a master oscillator for timing and synchronization. The power of the individual transmitters add together to give an acceptable output. Another advantage is that because there are no moving parts, a great deal of complicated mechanical equipment is eliminated, and the scanning rate is no longer limited by the inertia of a bulky antenna.

Pulse-Compression Filters

To increase the resolution of a radar system a narrow pulse is necessary, but unfortunately a narrow pulse requires greater power. More power is also needed for greater range. Ideally, we would like to transmit a wide pulse with relatively low power, but receive a narrow pulse with power high enough to make it stand out against background noise! This seemingly impossible task is accomplished by a technique known as **pulse compression**.

In this method a series of pulses—let's take four, for example—are transmitted at moderate power. Each has a different frequency. When the echoes come back, the four pulses are routed through a pulse-compression (or dispersive) filter. This filter consists of four delay lines, and the pulses are separated according to their frequencies. The first pulse travels through the line giving the most delay, the second through a line with less delay, and so on with the third and fourth pulses, with the result that they all arrive at the receiver simultaneously. Consequently, their individual amplitudes are added to form a single pulse with a total amplitude four times as great, but with the same width as each of the original pulses received by the antenna. In this way, we get the narrow higher-power pulse at the receiver, but we do not have to transmit one.

In an actual radar system using pulse compression, the number of pulses may be many more than the four used in the explanation

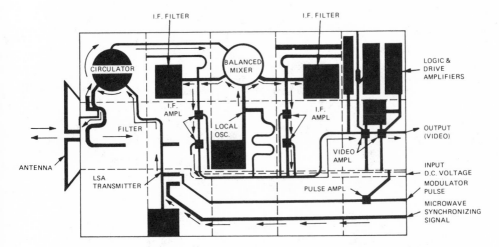

Figure 8.8 Microelectronic radar system (compare with Figure 8.1).

above. The number may even run into the hundreds, depending upon the mode of operation of the radar system.

Figure 8.8 is a block diagram of a microelectronic radar system containing the features so far discussed, and some others that we shall now consider.

Circulators

The first of these is the circulator. This device makes use of a property of ferrites which causes them to be unidirectional at microwave frequencies. A round patch of a ferrite material such as single-crystal yttrium iron garnet (YIG) is formed on the beryllium-oxide chip beneath a circular microstrip conductor. Microwave signals can travel only in one direction around the circulator because of the interaction between their magnetic field and the ferromagnetic properties of the ferrite. In Figure 8.9, energy from the transmitter enters the circulator at A and travels toward B. It cannot go in the opposite direction toward C. If B is unattenuated, practically all the energy leaves there to go to the antenna. Similarly, when an echo from the target is picked up by the antenna, it enters at B and is conducted to C, which is connected to the receiver.

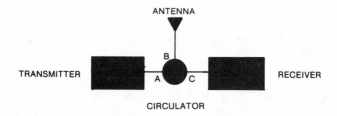

Figure 8.9 The place of the circulator in a microelectronic radar system.

Delay Lines

The pulse compression filter has already been mentioned. It is also called a dispersive filter or dispersive delay line. The pulses are separated according to their frequency by an interdigital structure with a non-uniform grating, as shown in Figure 8.10.

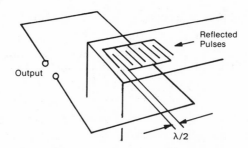

Figure 8.10 Microacoustic, interdigital structures with non-uniform grating can be used as dispersive delay lines.

First frequency out would be the one whose half wavelength corresponded to the spacing of the first interdigital pair accosted by the reflected acoustical wave. Second frequency out would be the one whose half wavelength matched the spacing of the next interdigital pair accosted by the reflected wave . . . etc. The spacing would not have to be varied in any particular sequence—thus, the variations could be used to set up pulse coding patterns.

The delay line is made of a ferrite material such as YIG. In this material the amount of delay varies with the frequency. You can see how this allows one short length of YIG to provide the different delay times required in the filter. The spacing of the pairs of elements in the structure illustrated in Figure 8.10 is arranged to equal half-wavelengths of the pulse frequencies, so that they select only those required. Other frequencies are greatly attenuated.

The local oscillator may be a Gunn-effect diode or a transistor with a frequency multiplier. Its signal beats with the antenna signal in the transistor mixer to produce the intermediate frequency. The local oscillator-mixer combination is often called the RF down converter, because that is what it does.

Standard radar IF frequencies are 60 or 30 megahertz, so the IF amplifier will not be much different from a microelectronics TV IF amplifier. The bandwidth required will depend upon the spread of frequencies involved in a pulse compression technique, and also upon the antenna system being used. The video detector and video amplifier are also similar to their TV counterparts.

Phase Shifters

Where phase-shifting is used (as with signals from phased-array antennas), PIN-diode switches are employed. A PIN diode is one that consists of a layer of semiconductor that is intrinsic, or relatively lightly doped with both p and n dopants, sandwiched between layers of heavily doped p and n material. ("PIN" is an abbreviation for p-type/intrinsic/n-type semiconductor diode.) With reverse bias the middle layer acts as an insulator, but with forward bias both outside layers inject carriers into the center layer, which then becomes a conductor. In this way, the diode acts as a switch. It is capable of handling high power in spite of its extremely small size, which makes it perfect for microelectronic radar.

The phase-shifting circuit is made up of a length of meandered microstrip (Figure 8.11) with dimensions chosen to suit the frequency of the signal and the phase angle required. This is switched into and out of the circuit by a master oscillator controlling the phased array. This type of phase shifter can also be made bidirectional so

Figure 8.11 Phase Shifter.

Ferrite phase shifters are available in waveguide, coaxial, stripline and microstrip types. The microstrip units can be reciprocal or nonreciprocal. For a phased array antenna, the reciprocal type can handle both transmitter (T) and receiver (R) functions without switching.

that it can handle signals for both transmitter and receiver without switching from one to the other.

The remaining circuits in a microelectronic radar system are similar to those already discussed for use in other equipment. Much of what we have covered in this chapter can be used for other types of microwave circuits. Pulse compression is of particular value for communication in outer space, where we may soon be beaming microwave signals over distances in the light-year range. (A light year is approximately 5.87×10^{12} miles.) High-power microwave equipment is also being developed for use in space stations which will gather solar radiation, convert it to microwave frequencies, and beam it down to earth.

Consumer Microwaves

Microelectronic microwave equipment was originally developed for military and space uses, with the taxpayers footing the bill for all of the early experimental work. Now those same taxpayers are in a position to get some of their money back, since the bread they cast upon the waters has returned in the form of affordable microwave devices they can use for their own purposes. These include direct reception of satellite broadcasts, some types of burglar alarms, "intelligent" microwave ovens, radar in automobiles that tells you how far you are behind the car in front, and so on.

Satellite reception by amateurs has involved fairly large dish antennas (about 12 feet in circumference for 3.7 meters) because of the frequencies in use by existing "birds." However, the proposed pay-TV satellite system at a higher frequency (12 gigahertz) will require only 2.5-foot dishes mounted like ordinary TV antennas. (These will probably be the next "status symbol.") In the 3-meter system, a **low-noise amplifier (LNA)** and **down-converter** are mounted with the **feedhorn** at the focal point of the antenna dish. This assembly converts the gigahertz satellite signal to a megahertz signal for delivery by ordinary coaxial cable to the indoor receiver. The latter contains the tuning circuits for selecting the satellite channel, and for converting the megahertz signal to that of one of the TV channels used in the standard TV receiver.

Intelligent microwave ovens still use magnetrons to generate the microwave energy, since this is quite considerable (input power of around a kilowatt is typical). However, the "computerized" control circuitry which asks the cook what he or she wants to do, and then does it automatically, is similar to that already described in earlier chapters.

Automobile radar is an option that is made possible by the developments described in this chapter. It informs the driver of obstructions in his or her path, mainly other vehicles, so that he or she can avoid coming too close, especially under reduced visibility. Connected to the "cruise control," it can maintain a minimum safe distance.

This use of forward-directed radar is complemented by a rearward-directed sonar device, which emits an ultrasonic pulse and measures the time interval between its emission and the receipt of its echo. This is not radar, of course, because it makes use of the focusing system introduced in the Polaroid camera. A driver backing up his vehicle to, say, a loading dock, can know exactly how much space there is between him and the obstruction, even if he cannot see it at all.

These examples illustrate some uses of microwaves available to the consumer that have been made possible by the microminiaturization of microwave circuits. They are low-power devices, that with normal usage present no hazard to people. (The type of radiation emitted at microwave frequencies is, of course, totally different from that of radioactive particles; as is well known, meals cooked in a microwave oven do not become radioactive.)

CHAPTER **9**

The Optoelectronics Connection

Optical Communications

The continuing rise in the cost of copper and the ongoing broadening of communications bandwidths make alternatives to communication by ordinary copper wire more and more attractive. (See Table 9.1 for a comparison of the various transmission media.) Satellite and microwave links have, for some time, been carrying much of the traffic that previously would have been borne by copper. Now, fiber optics provides a transmission medium that is superior in most respects to all other media, and that permits the realization of such systems as the **Integrated Services Digital Network (ISDN)**, which integrates voice, data, and video signals all on one fiber. ISDN is al-

Transmission Type	Frequency Range	Attenuation
Wire pair	1–140 kHz	0.1–0.3 dB/km
Coaxial Cable	0.06–51 MHz	59 dB/km*
Waveguide	≤ 50 GHz	0.5–4 dB/km
Optical Fiber	10^{14} Hz	< 1.0–800 dB/km
*RG 59/U at 28 MHz		

Table 9.1 Comparison of Various Means of Transmission.

ready in operation in Europe, supplying telephone, television, and radio services and, in some cases, even mail service to the subscriber.

Optical Spectrum

Any communication making use of that portion of the electromagnetic spectrum between 10^1 and 10^6 nanometers is by convention an optical communication, even though visible radiation is quite a small segment of this portion, which includes ultraviolet (UV) and infrared (IR) wavelengths as well. The subdivisions of this band are as follows:

10–200 nm	Extreme UV
200–300 nm	Far UV
300–370 nm	Near UV
370–750 nm	Visible light
750–1500 nm	Near IR
1600–6000 nm	Middle IR
6 100–40 000 nm	Far IR
41 000–10^6 nm	Far-far IR

Practical considerations do not allow use of all of this enormous range of wavelengths. The bands of major interest in communications are 800–900 nm and 1300 nm. The longer wavelength has the advantage that cable losses are less than 1 dB/km, compared to 6 dB/km at 800 nm.

Fiber Optics

Communications-systems engineers have known the advantages of fiber optics for some time: more bandwidth than conventional systems; compact size; absence of crosstalk and electromagnetic interference; much less susceptibility to heat and caustic environments; no fires and explosions when a cable is broken. But the difficulties that had to be overcome have been formidable. Chief among these were the development of components operating in the 1300-nm region to take advantage of the lower cable loss at this wavelength. These problems are discussed later under the sections dealing with sources and detectors.

Optical-fiber waveguides can be made of fused silica, glass, or transparent plastic. Because fused silica fibers are now available with very low attenuation, this material is preferred. The fibers have a diameter of as little as 0.05 mm, with an outer covering called a cladding. The cladding is there to protect the high polish on the fiber to maintain total internal reflection. There are three types of fiber: single mode, multimode, and graded refractive index.

The **single-mode fiber** is one in which its *normalized frequency* (V) does not exceed 2.405. V, which varies directly with fiber radius and indirectly with wavelength, is given by:

$$V = 2\pi a(n_1^2 - n_2^2)^{1/2}/\lambda$$

where a = the radius of the fiber (see Figure 9.1).

n_1 = the refractive index of the glass

n_2 = the refractive index of the cladding

λ = the wavelength

A **multimode fiber** is one in which, generally, the value of 2.405 for V is exceeded. This results in the light taking more than one path through the fiber (see Figure 9.2), so that pulse blurring takes place. If the data transmission rate is high, adjacent pulses tend to merge as shown in Figure 9.3. This limits the maximum data rate the fiber can handle. The number of modes (N) is given by:

$$N = V^2/2$$

The **graded-refractive-index fiber** is made with material with a refractive index that changes with a parabolical profile (see Figure 9.4) from center axis to outer surface. As light rays deviate from the center axis, they are bent back toward it. At the same time their velocities are equalized so that pulses are much less distorted.

Because of the pulse-broadening problem, the multimode fiber is not suitable for the transmission of *pulse-coded modulation*

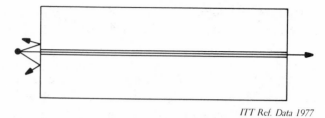

ITT Ref. Data 1977

Figure 9.1 Propagation in Single-Mode Fiber.

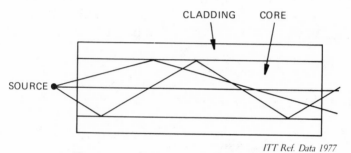

ITT Ref. Data 1977

Figure 9.2 Propagation in Multimode Fiber.

DISTORTED PULSES

ITT Ref. Data 1977

Figure 9.3 Pulse Broadening in Multimode Fiber.

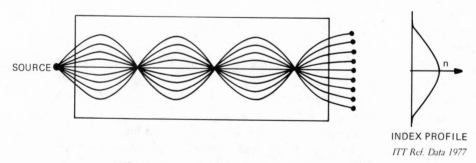

INDEX PROFILE

ITT Ref. Data 1977

Figure 9.4 Graded-Refractive-Index Fiber Propagation.

(PCM). Another code system, called *delta modulation* (DM), has to be used. In this method positive and negative pulses indicate the direction of change of modulation signal amplitude. (See Figure 9.5.)

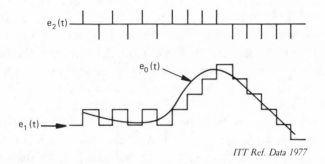

ITT Ref. Data 1977

Figure 9.5 Delta Modulation. Original audio signal e_o (t) is digitized into positive and negative pulses e_2 (t). When these are demodulated and integrated e_1 (t), they reconstruct the original audio signal.

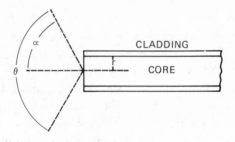

Figure 9.6 Maximum Acceptance Angle.

Figure 9.6 shows how a fiber conducts a ray of light so that it goes around corners. In this fiber, the low-loss core has a refractive index n_1 and the cladding a refractive index n_2. Light must enter the end of the fiber at an angle that does not exceed the maximum acceptance angle (θ), which is given by:

$$\sin \alpha = (n_1^2 - n_2^2)^{1/2}$$
$$\theta = 2\alpha$$

As long as the angle that a ray of light makes with the internal surface of the fiber does not exceed this value, it will be internally reflected. The fiber can therefore be bent up to this limit without allowing any light to escape.

The value sin α is also called the **numerical aperture (NA)** of the fiber. Fibers with core diameters between 50 μm and 200 μm have NAs from 0.2 to 0.3. The smaller diameter permits a higher bandwidth, but the larger NA allows more light to enter the fiber.

Fiber-optics waveguides are not the same as the *flexiscopes* used in medicine and industry for the investigation of otherwise inaccessible sites. These use fibers with diameters down to 5 μm, packed in bundles of several hundreds or even thousands. A *fiber-optics cable* with 144 fibers packaged in a polyethylene sheath is about one-half inch in diameter, and carries the equivalent of more than 40,000 voice channels.

Lasers

Solid-state lasers and light-emitting diodes (LEDs) are the sources that supply the signals carried by optical fibers. Lasers are the more powerful. They also are more suitable for single-mode operation, because of their narrow bandwidth. However, they are considerably more expensive and do not have as long a life as LEDs.

Gallium arsenide (GaAs) is the semiconductor that works best in a laser designed for the 800–900 nm band, but is not suitable for use at 1300 nm. For the latter wavelength, the semiconductor used is one combining indium, gallium, arsenic, and phosphorus (InGaAsP). As shown in Figure 9.7, a PN junction is formed between regions of p-type and n-type semiconductor. When the junction in a solid-state laser is forward biased, large numbers of the atoms, ions, or molecules of the semiconductor are in an excited state. The minority carriers recombine at the junction, emitting photons as they do so. When more atoms, ions, or molecules of the semiconductor are in an excited state than in an unexcited state, optical gain exceeds optical loss, and stimulated emission occurs. This is because photons emitted by the excited entities as they return to the ground state stimulate the emission of still other photons.

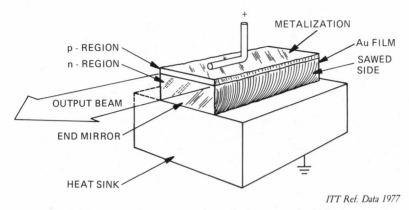

Figure 9.7   Injection Laser. The end mirrors are of gold film.

Facing mirrors at either end of the active region provide a reso-
nant cavity in which regenerative gain takes place for photons traveling coaxially along the cavity axis. The dimensions of the cavity are
chosen so that one or more of its resonances falls within the laser
transitions for which gain exceeds loss. This positive feedback results
in oscillation at frequencies which vary with the semiconductor material.

One of the mirrors is 100 percent reflective, the other about 95
percent. This allows light to emerge from the latter in a narrow
beam.

This type of laser, which is called an injection laser, is characterized by a very fast risetime (about 1 ns). It can deliver about 10
mW over a spectral width of about 1 nm, and can be both pulse-
and amplitude-modulated, using direct modulation, which is described further on.

Coupling the output beam from the laser into the optical fiber
requires that the axes of the fiber and the laser cavity are in line, so
that the laser beam makes an angle with the fiber axis that is within
the acceptance angle of the fiber.

Light-Emitting Diodes

Figure 9.8 shows the structure typically used for a LED
designed as a source for optical fiber signals. It consists of a

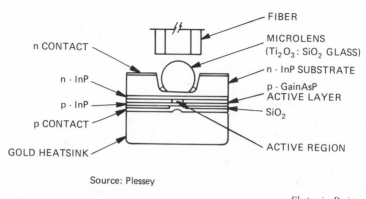

Source: Plessey

Electronics Design

Figure 9.8 Light-Emitting Diode with Microlens.

Dagwood sandwich of several layers. The active layer is of the same semiconductor materials as in the laser just described. It is placed between two *confining layers* to keep the light energy from dispersing in the source itself. Only the middle portion of the active layer emits light. Photons are released by recombination when the junction is forward biased by current flowing between the ohmic contacts. The photons emerge from this region and are collimated by the microlens so that the maximum number can enter the fiber.

The wavelength of the emitted light depends upon the transition time between the higher and lower energy levels. The width of the energy gap is dependent upon the composition of the elements used in the active area. GaAs (or GaAsAl) produces light with a wavelength below 900 nm. InGaAsP produces light between 1300 and 1500 nm. Exact frequency is determined by the composition and dimensions of the active region. Fabrication of the chip follows lines similar to those for transistors, except that silicon is not used.

PIN Diodes

The most satisfactory type of diode for changing photons back into electrons is the PIN diode. When fabricated of InGaAsP, it operates satisfactorily in the 1300-nm region, and with GaAs, or GaAsAl material, it does equally well in the shorter wavelengths.

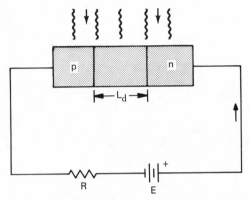

From Technician's Guide to Solid-State Electronics, *Morris Grossman,* Parker, 1976, page 126.

Figure 9.9 PIN Diode. L_d is the "i" region.

Figure 9.9 shows how a PIN diode functions. Between a heavily doped n region and an equally heavily doped p region is sandwiched a layer of intrinsic, or only very lightly doped, semiconductor material. For operation in what is called the conduction mode, the diode is reverse biased. When the junction is reverse biased the "i" layer has a high resistivity, and most of the applied voltage appears across it. Light falling on the "i" layer generates hole-electron pairs. These would recombine pretty fast, but because the width of the layer is narrow and the attraction of the p and n regions strong, most of them are collected by the p and n regions. The resultant current is proportionate to the strength of the illumination applied to the i region. These diodes have a very fast response (in some cases less than 1 ns) to pulsed light, so that they are very suitable for data detection. However, they can also detect analog signals, since the diode current varies with the intensity of the light.

Modulation and Demodulation

As mentioned previously, both injection lasers and LEDs can be modulated with direct modulation. This means that the modulating signal is made to vary the bias applied to the device. This makes for very simple circuitry, as shown in the example in Figure 9.10.

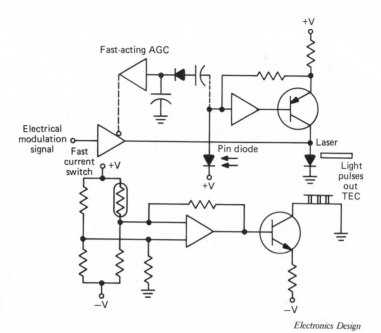

Electronics Design

Figure 9.10 Modulation Circuit for Laser. Note thermoelectric cooler (TEC).

As you can see, one of the main considerations in this circuit is temperature control. The wrong temperature not only cuts operating hours from the life of the laser, but also affects its operating characteristics. To prevent output power from varying uncontrollably, a special circuit is required to stabilize the output. Thermoelectric coolers keep the junction temperature of the laser at a constant 60°C. A PIN diode also senses any change in the output light and applies an appropriate correction to the bias current.

For demodulation, it is usual to provide an integral preamplifier as shown in Figure 9.11. The preamplifier increases the sensitivity of the PIN diode and provides noise reduction as well. The output signal is thus prepared for application to a main amplifier. In this IC an additional identical PIN diode that is not exposed to light from the fiber is provided to generate dark current only. This is subtracted from the output of the exposed diode, so that only the signal appears at the output.

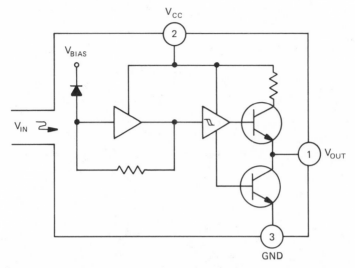

Electronics Design

Figure 9.11 Demodulation Circuit for PIN Diode. The internal PIN diode generates dark current only to buck that of the exposed PIN diode.

Integrated Services Digital Network

As mentioned at the beginning of this chapter, ISDN is already in operation in Europe. In many countries all communications are supplied by a single source, usually the post office. In addition to mail, that agency also supplies telephone, television, and radio. This avoids the situation in the United States, where a number of independent commercial organizations have to get together and agree on a common system before a start can be made.

In the U.S., ISDN will be handled by the telephone companies. As shown in Figure 9.12, a subscriber can be provided with voice, data, stereo and video services, giving him telephone, videotex or other data service, music, picturephone and color TV, all via one optical fiber. The one optical fiber actually serves four subscribers, each with its own frequency. On a convenient telephone pole or buried in the ground, close to the subscribers' residences, is a passive device called a wavelength-division multiplexer (WDM) that filters out each of the four frequencies for transmission to the four subscribers.

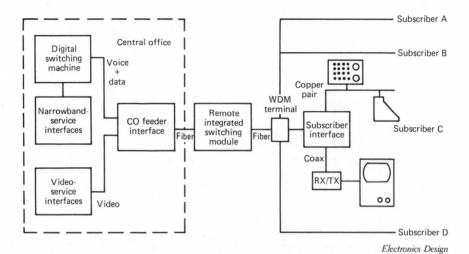

Electronics Design

Figure 9.12 Integrated Services Digital Network. In this system, each of the four subscribers can have, among other terminals, a data set, a telephone, a video telephone, a color TV, and a stereo receiver. Four different wavelengths are placed on the cable by the central office feeder interface. At some distant location, a remote integrated switching module and WDM separate out each wavelength and distribute the services to the four subscribers.

In this way, the telephone company does not need to have a separate optical fiber for each subscriber except for the short distance between the WDM and the residence. In the current state of the art, four frequencies is the maximum that can share one fiber, since lasers with more precision would be too expensive, and it would be equally costly to provide filters (e.g., prisms) sharp enough to select the proper signal. This system is designed to be competitive with other vendors, each of whom is using a copper wire cable supplying only one of the services available from ISDN.

ISDN is particularly valuable to offices because of the various uses to which it can be put (e.g., electronic mail). However, unlike existing coaxial networks such as Ethernet, it is also attractive for home use because of the entertainment and convenience of the services available.

Displaying Information

The main types of displays used with solid-state equipment are LEDs, LCDs, vacuum-fluorescent and gas-discharge displays, and cathode-ray tubes (CRTs). The type selected for a particular application will depend upon the type of information to be displayed, the reading distance, power availability, environmental requirements, and cost.

The "character fonts" that may be used are shown in Figure 9.13. Most numeric applications (e.g., digital clocks and watches, calculators) will not need more than a seven-segment display. However, where some alphabetical upper-case characters are also required, a 14- or 16-segment "starburst" may be used. For full alphanumeric capability, with upper- and lower-case letters and some special symbols, a dot-matrix font must be used.

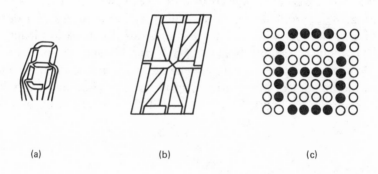

(a)   (b)   (c)

Figure 9.13 Solid-State Displays: (a) the connections are transparent in the 7-segment display; (b) a starburst display with 16 segments; (c) a dot-matric display.

The reading distance determines the character size. For a handheld device, such as a calculator, the comfortable viewing distance is 0.6 m. The typical character height is 2.5 mm, although larger character sizes are often provided. For a desktop computer, the viewing distance is from 0.9 to 1.2 m, with a minimum character height of 4 to 5 mm. A benchtop instrument with a viewing distance of 1.9 to 2.7 m should have a minimum character height between 7.5 and 11 mm. Appliances with viewing distances from 3.5 through 6 m will need character heights from 14 mm through 25 mm.

LEDs

As you have already seen, LEDs can emit photons in both the visible and infrared parts of the electromagnetic spectrum. A lot depends upon the composition of the semiconductor. This in turn determines the energy gap between the valence band and the conduction band. If the material allows the recombination of electrons and holes directly across the energy gap, photons are given off with energies that are restricted to a narrow band of frequencies corresponding to the energy gap potential (E_g). In gallium arsenide phosphide (GaAsP), E_g varies according to the proportion of As to P, so that both green and red LEDs can be obtained.

In the segmented displays used in calculators and similar devices, tiny LEDs are arranged to form the figures and letters by activating the appropriate segments. This is done by means of a display driver or decoder which, you will recall, was discussed in Chapter 3. LEDs require a supply from 2 to 8 V and consume from 1 to 200 mW power per **pixel** ("picture element"), depending upon their size and the amount of information displayed. They also have a very long life, typically 10^7 hours. Character height varies between 2.5 and 50.8 mm.

Liquid Crystal Display (LCD)

A liquid crystal is a material in a state halfway between liquid and solid. The type favored for most LCDs is called **twisted nematic**. A very thin layer of this material, about 12 μm thick, is sandwiched between two glass plates. Its molecules normally lie in layers parallel to the glass plates. All the molecules in *each* layer are oriented in the same direction, but successive layers are rotated through increasing angles until the molecules in the bottom layer lie approximately at right angles to those in the top layer. This twisted structure causes a ray of light passing through the LCD to turn on its axis through 90 degrees.

Each glass plate has a thin layer of polarizing material deposited on it. These are angled so that if there were no liquid crystal between them they would block the passage of light. However, the rotation of the light caused by the liquid-crystal molecular structure neutralizes this, and so the glass-liquid crystal-glass sandwich appears clear.

But if an electric field with transverse lines of force is applied to the liquid crystal, the molecules influenced by the field rearrange themselves parallel to the lines of force and their twisted structure disappears. Now, light passing through the liquid crystal is not rotated, so it is blocked by the polarizing filters, and the affected area appears black.

In the LCD, this field is applied between transparent tin-oxide electrodes printed on the front and back plates. The electrodes on the front plate can be in any desired form, and mostly consist of a figure eight broken into seven segments. Each segment is connected to one of the seven outputs of the display driver, and one or more are energized according to the binary input to the ROM decoder, as shown in the truth table in Figure 3.16. The backplate electrode is the common electrode and is energized all the time. Obviously, there must be a separate decoder for each figure in the readout unless a multiplexer is also provided to switch the figures in rapid succession.

An LCD readout is activated by a square wave with a frequency between 30 and 200 hertz. (If DC were used as in other types of display, the liquid crystal would not clear fast enough after removing the voltage.) The frequency is not critical, but if it is lower than 30 hertz, a flicker would be evident in the display. Too high a frequency, on the other hand, would not allow enough time for the molecules to realign themselves between cycles.

The square wave is applied to the DF IN terminal (DF = display frequency). Its amplitude is not critical because it will be adjusted by the level shifters. The DF OUT signal is applied to the common electrode. Square-wave signals from the display driver that are applied to the segments to be energized have a polarity opposite to the DF OUT signal, so the voltage across the display is doubled. Unactivated segments, on the other hand, are supplied with an in-phase square wave, so the effective voltage across these is zero (see Figure 9.14.)

LCDs require a supply from 3 to 10 V and consume from 0.3

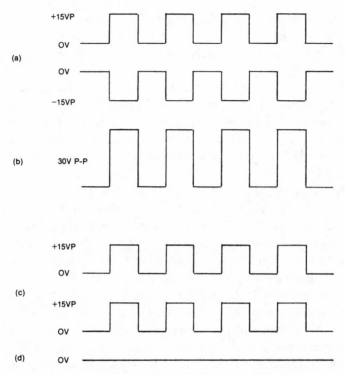

Figure 9.14 Square Wave Phasing for LCD. Square waves of (a) are of opposite phase, so voltage difference (b) is their sum. Square waves of (c) are of similar phase, so voltage difference (d) is zero.

to 100 mW power per pixel. Character height varies between 2.5 and 50.8 mm.

Vacuum-Fluorescent Display

This is a vacuum tube in which the anode is made up of a segmental display. Electrons from the cathode are attracted to those segments that are activated by application of the anode voltage. Each segment is coated with a fluorescent substance which glows when the electrons strike it. The anode requires from 12 to 60 V, and the filament from 1 to 12 V. Power consumed is from 1 to 100 mW per pixel. Character height varies between 4.2 and 30.0 mm. This is a bright display, visible in high ambient light conditions.

Gas Discharge Display

As in the fluorescent display, the gas discharge display is a vacuum device in which electrons emitted by a common cathode strike whichever segment (or dot, in a dot-matrix display) is energized by the anode voltage. An orange-red glow appears at the positive electrode as in a neon bulb. The display operates with a supply voltage of 130–250 V and a power consumption of 3–250 mW per pixel. The character height is between 5.1 and 88.0 mm.

Cathode-Ray Tube

As you know, a CRT operates by means of an electron beam generated in an electron gun and directed at a phosphorescent screen. The point at which the beam strikes the screen is determined by the potentials upon deflection plates within the tube (electrostatic deflection) or upon the currents in deflection coils around the outside of the neck of the tube (magnetic deflection). The phosphor screen of the tube is made to glow by the impact of the electrons striking it.

Various phosphors are available for producing different colors, so that the display can be black-and-white, black-and-green, colored, and so on. Character height can be from 2.5 mm up to the largest size the screen can accommodate. Much more information can be displayed on this screen than on any other readout device. The fonts that can be used are highly flexible. The supply voltage for the anode is from 5 to 25 kilovolts.

Solid-state display screens have not yet replaced the CRT, and will not do so until they can be made as efficiently and as economically. The principle, however, is well understood, and has been demonstrated experimentally. Figure 9.15 illustrates the kind of circuitry involved. The screen consists of 100 LEDs connected to intersecting lines as shown. A 4017 IC acts as a horizontal sweep by energizing the vertical lines in turn from left to right, and an LM3914 dot/bar display driver energizes one or more horizontal lines according to the input voltage. Where two energized lines intersect, the LED at that point will light. This particular "CRT" operates like that in an oscil-

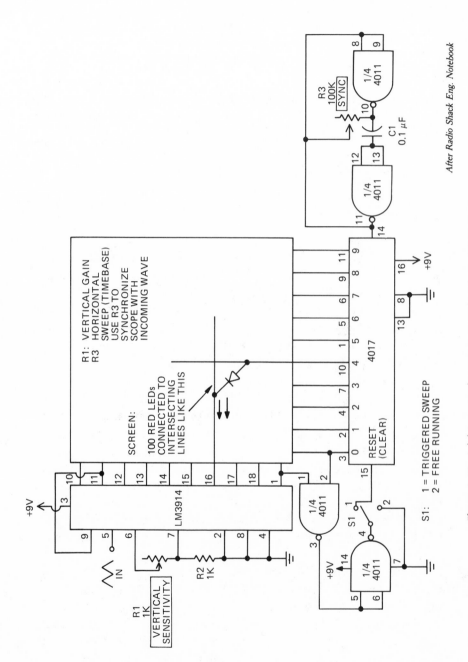

Figure 9.15 Solid-State Display Screen: "picture tube" of the future?

loscope, and obviously has very poor resolution. For TV, both horizontal and vertical sets of lines must be energized sequentially at the appropriate rates, while the brightness of each LED will depend upon the input signal amplitude. Color TV involves the provision of three sets of LEDs with the primary colors of red, green and blue, and the manufacturing process requires that hundreds of thousands of them, with their interconnections, all be microelectronically fabricated on a suitable substrate. You can see that to get such a process going needs a huge investment, so that as long as the vacuum-tube CRT continues to perform satisfactorily it will go on being used. The solid-state screen may have to wait for military necessity to bring about its development.

Optoisolators

An optoisolator, or optocoupler, which might be thought of as a solid-state isolation transformer, consists of a light source and a light sensor, packaged together so that the radiation from the source is coupled, usually directly, to the sensor. Electrical paths between the source and the sensor are carefully avoided so that signals applied to the input couple only optically to the output. A material such as optical glass, which is good insulator, is placed between the source and the sensor. This enables the distance to be kept small to achieve the maximum current-transfer ratio (CTR). However, some leakage resistance and coupling capacitance always exist. The resistance value is on the order of 10^{11} ohms or higher, and the capacitance does not exceed 3 picofarads. This is an excellent level of isolation for most applications, and permits separation of widely different voltage levels.

Most optoisolators today employ an infrared LED for the source, and a photodiode or phototransistor for the sensor. The sensor must have a response that is maximal for the emission wavelength of the source. Phototransistors are used more than diodes since they have a higher output. When diodes are used, they usually have a transistor packaged with them to amplify their weaker signals. Some optoisolators incorporate Darlingtons, SCRs or triacs in their output.

The most popular form of an optoisolator is a dual-in-line package (DIP) with six pins. Figure 9.16 shows the basic circuit for one of these.

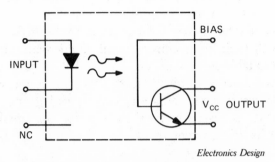

Electronics Design

Figure 9.16   Optoisolator.

CHAPTER 10

Making Printed Circuit Boards Is Easy and Fun

The evolution of the printed circuit board (PCB) from the handwired chassis began at the end of World War II, at which time the bipolar transistor had not been invented, and integrated circuits were not even dreamed of. Yet, its seemingly teleological evolution culminated in time to provide the perfect means for mounting and interconnecting the ICs and other featherweight components of the spage age.

The PCB is not only lightweight and durable; it is reliable as well. Reliability is all-important in devices that must function in harsh environments where maintenance or repair are impossible. The amazing performances of unmanned probes to other parts of the solar system are proof of the incredible dependability of modern electronic circuitry using PCBs, especially when you consider that the failure of just one simple solder joint among the multitude of connections in that equipment could abort a mission costing millions. (See Figure 10.1, also Figures 11.2 and 11.3.)

Types of Boards

These are the main questions involved when choosing a PCB:

Copper-clad on one or both sides?

Paper-based or glass-based laminate?
Thickness of laminate?
Thickness of copper?
Type of input/output connectors?

Factors Affecting Choice

If the circuit is simple, and if there are few if any crossovers contemplated, then the obvious conclusion will be that a single-sided board will serve the purpose. On the other hand, in a situation where high-density packaging is a must, then a double-sided board may have to be used. It goes without saying that a double-sided board is more complex and therefore would be more costly to manufacture, whereas a single-sided board not only is less costly but also is easier to design and manufacture and less subject to reject. For these reasons (among others), most high-quality PCBs for space electronics are single-sided.

The most common PCB laminates fall into two broad categories, those with phenolic bases and those with epoxy bases. Phenolic boards are almost always fabricated with paper filler, while epoxy boards may be fabricated with paper or glass-cloth fillers. A third resin, teflon, is also available with a glass-cloth filler. Phenolic boards (NEMA grade XXXP) are the most widely used because of their low cost and are found in TV, radio, computer, and automotive circuits. Epoxy paper boards have similar applications. It is easy to punch holes in both types which is important from a manufacturing point of view. They do absorb moisture, however, if used in a damp environment, but on the other hand, epoxy boards with glass-cloth filler can withstand higher temperatures, are flame-retardant, practically impervious to moisture, but are sometimes not suitable for punching. Teflon boards are best for ulta-high frequencies because of their low dielectric losses.

These boards are produced in thicknesses from $1/64$ to $1/2$ inch. The thinner sizes should be used only for boards of small area, since larger boards would have a tendency to warp or crack. A thickness of $1/16$ inch is suitable for most projects.

The copper-foil plating is supplied in two thicknesses, "1 oz."

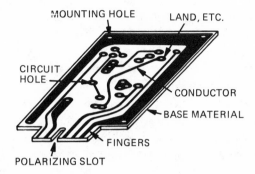

MOUNTING HOLE   LAND, ETC.

CIRCUIT
HOLE

CONDUCTOR

BASE MATERIAL

FINGERS

POLARIZING SLOT

Figure 10.1 Printed Circuit Board (solder side). Note "fingers" (male connectors) for edge connector.

(0.0014 inch), and "2 oz." (0.0028 inch). The 1 oz. is the one generally selected for most purposes.

Card-edge connectors are used on many boards. In other words, the male connectors are printed on the edge of the board that is inserted into the female socket, and are part of the printed circuit. The thickness of the board must, of course, be such that the male connectors make good contact. Poor connections are the most frequent cause of trouble in equipment using PCBs, so the importance of good connectors cannot be overstressed. Gold-plated contacts are often used.

Manufacturing Procedures for PCBs

Before fabrication can begin, the master artwork must be prepared. This has to be done with care because components for use on PCBs in many cases have leads spaced to match a standard grid. Initially, a blue-line gridded Mylar sheet is set up on the drafting table. The grid lines are generally on 0.1-inch centers. However, the drawing is two to five times oversize. The draftsman uses black pressure-sensitive tape for the conductor pattern with adhesive black disks for the terminals, or draws the design with opaque, permanent black ink. This master is then photographed to produce a negative the same size as the board.

This negative can be used in various ways. Best results are

obtained by exposing a sensitized board to light (often ultraviolet) through the negative. A sensitized board is one that has a coating of photo resist. The conductor pattern on the negative is white, of course, and the photo resist is exposed only where illuminated by the pattern. The exposed portion becomes insoluble, so that when the board is treated with a solvent, such as trichloroethylene, this portion remains in place while the rest of the photo resist is washed away. The board is then placed in an etching bath containing a mordant such as ferric chloride, which eats away all of the unprotected metal. After the etching has been completed, the remaining photo resist is removed with a stripping agent.

Because photoetching is rather slow and requires a darkroom (or special lighting), the silkscreen process is favored for work not requiring high resolution. The silkscreen is made in a similar manner to the negative described in the previous paragraph, but once made, it can be used to fabricate a large number of PCBs with greater simplicity. The screen is mounted in a frame, and as each sensitized board is placed under it, an acid-resist ink or lacquer is squeegeed through it on to the copper. After this dries, the board is placed in the etching bath as in the photoetching process.

Etching away the copper is called a *subtractive* process. An opposite process, called *additive*, is a plating process in which the circuit pattern is printed with conductive ink on an unclad board. The conductor pattern is then deposited in a manner similar to electroplating. The increasing cost of copper makes this method more and more attractive. It also has the advantage that if the holes are punched or drilled first, they can be "plated through" for making connections to the other side of the board. Other additive processes include metal spraying and die stamping.

As the boards may have to be stored for a while before they are used, some method of finishing is necessary to protect the conductor pattern. The most widely used method is to dip or wave-solder the boards so that the copper is coated with a layer of solder. This also makes subsequent soldering of parts easier. In some special applications silver plating is used, or solder may be applied by hot rolling or plating. Plating with gold, or rhodium and nickel, is generally used only for high-reliability PCBs used in military or space equipment. These boards are also treated later with acrylic or other insulating coatings to resist humidity and to increase the firmness of parts attachment.

Holes for component leads are punched or drilled. As already mentioned, some boards must be drilled because they cannot be punched. Special drill bits that leave a smooth hole must be used if holes are to be plated through. Sometimes eyelets or tubelets are inserted in the holes. The diameters should be not less than ⅔ the thickness of the board, and should not exceed by more than 0.020 inch the diameter of the lead to be inserted.

All components are mounted on the side opposite the conductor pattern, since the latter is to be dip- or wave-soldered. Any component which must be mounted on the conductor side has to be hand-mounted afterward, so this should be avoided if at all possible. In mass production, components are mounted by machines that bend the leads, cut them to the proper lengths, and insert the leads into the holes. They then crimp the lead wires protruding from the conductor side to prevent the parts from falling out before they are soldered in place.

Dip-soldering consists of applying flux to the conductor pattern, and placing the board in contact with 60/40 tin-lead solder at a temperature of 230° C. (450° F.) for five seconds while slightly agitating it. Wave-soldering is a refinement where, instead of placing the board in contact with the solder, it is passed horizontally over the surface of the bath. A pump creates transversely across the bath, a one-inch high ridge or hump that wets the surface of the PCB as it passes over it. The advantages of this method are that the entire board is not exposed to maximum heat all the time; the temperature of the solder is not lowered where it is in contact with the board; and oxidation of the solder is prevented.

Solder masks are frequently used for these types of soldering to restrict the areas of the board wetted by the solder. This prevents "bridging" between closely spaced conductors, and concentrates the heat of the solder around the joints to be soldered. The mask material is applied with a silkscreen in the same way as acid resist.

Making a PCB

Making your own PCB is quite easy, inexpensive, and very in-

structive. All the materials you need are available at your neighbor-
hood electronic parts store.

The first step is to design the layout of the board. For this, the
most generally used procedure amongst experimenters and hobbyists
is to take a piece of paper the same size as the PCB, and position all
the components on it. This is so you can mark their terminal points.
Components such as resistors that have pigtail leads must have their
leads correctly bent first.

Then, with a pin or sharp-pointed pencil, indent the terminal
points so that you can see them plainly on the other side of the paper when you turn it over. This is done because the circuit has to go
on the opposite side of the board to the components. Now, draw the
connections between the terminals lightly with a pencil, avoiding
crossovers. Where a connection is to a component that is not
mounted on the board, such as a panel meter, the connection should
be drawn to a convenient point to which a hookup wire can be connected.

When you are satisfied that you have got all the connections
sketched in correctly, you can, if you wish, place small adhesive circular disks over the terminal points (where holes will be drilled) and
stick tape over the pencilled connections. This will give you a professional looking printed circuit template. However, the essential thing
is to obtain a clear PCB design.

Now, prepare the copper surface of the PCB for etching. Some
boards come with a transparent film over them that must be peeled
off. Others have a lacquer coating which must be removed by a *light*
scouring with fine steel wool. The etchant cannot dissolve all the
copper that has to be removed if the surface is not clean and bright.

Get a piece of carbon paper the same size as your PCB. Tape
it and the design to the board so they won't slip, and trace the circuit on to the copper surface with a ballpoint pen. Check the diagram to make sure all lines have been drawn. Then use a center
punch to indent the copper surface where the holes are to be drilled.

Now, apply the **resist** to the board. The simplest way of applying it is by means of a pen (obtainable from Radio Shack or some
other supplier, either separately or as part of a PCB kit). Make sure
the "ink" is emerging from the pen (it makes a black line), then
trace over the lines on the copper surface, making heavy dots where
the holes are to be drilled. When the resist has dried (after about a

minute), go over the lines and dots again to ensure complete coverage.

The **etchant** is a solution of ferric chloride, which is an orange-colored liquid that can stain your skin (and anything else it comes in contact with). It is also **poisonous** and **corrosive**, so you shouldn't drink it or breathe the fumes. It would be wise to wear rubber gloves and an apron, and perform the etching in a well-ventilated area.

You need a photographic developing dish a little larger than the PCB.* Put the PCB in it, and cover it with etchant to a depth of ¼ inch. Agitate the liquid back and forth, as you would in developing a photographic print, until the copper is all etched away from the parts of the board not protected with resist. This will take from 15 to 20 minutes. When all the unwanted copper has been dissolved, remove the board with tongs, pliers, or rubber gloves, and hold it under the cold water faucet for about two minutes. This will wash off the etchant and stop the chemical action. Don't keep the used etchant; it isn't any good for further etching.

The next step is to remove the resist, which is done with **resist ink solvent**. This is toluene, which is even worse than the etchant, since it is also **flammable**. Don't use the same dish you used for etching, and remove the final traces by scrubbing with household cleanser or fine steel wool under running water until the copper is bright and shiny. Use care in doing this, because the copper is extremely thin (0.0014 inch, remember?), and can be severely damaged with rough handling.

Your PCB is now ready for mounting the components. The first step is to drill holes through the dots so that the pigtail leads can be inserted. Use a ¹/₁₆-inch drill bit. Then drill a hole in each of the four corners of the board for mounting it. Use a ³/₁₆-inch drill bit.

The components are mounted on the opposite side of the board from the copper circuit. The leads come through the holes, and are soldered to the copper conductors, using a small iron (excess heat will cause the copper to lift from the board). Protect heat-sensitive components (e.g., transistors) by attaching a heat sink, such as an alligator clip or hemostat, to the lead being soldered. Make sure the tip of your iron is firmly seated in the heating element, and keep

---

*You will not need this if you purchase a Radio Shack PCB kit.

it clean by frequent wiping. Use the least amount of solder necessary, but be sure it flows and "wets" the lead and copper conductor properly. Also, solder into their holes the leads required to connect the board to the components that are not mounted on the board.

Maximum soldering temperature must not exceed 275° C., and must be applied for not longer than five seconds at a distance not less than three millimeters from the plastic case of the IC. Needless to say, only rosin or activated-rosin fluxes can be used in soldering electronic equipment.

Instead of mounting ICs directly on the board, they may be plugged into IC sockets previously soldered on to it. This avoids applying heat to the ICs themselves, and makes it easier to remove and replace them afterwards. The additional expense involved restricts this practice to more costly or critical equipment, and is mostly used for dual in-line packages (DIPs). Flatpacks and TO cans are nearly always soldered directly to the board.

In addition to avoiding damage from heat, you must be careful not to expose ICs, MOSFETs and high-frequency bipolar devices to electrostatic discharge. This is the discharge of the static voltage stored in your body's capacitance, and can occur when you pick up the device. For this reason, these devices are stored prior to assembly with their leads inserted into some conductive material (*not* polystyrene "snow," which is an insulator). When handling them, your hand should be grounded by a metallic wristband connected to a good ground. If a soldering iron is being used, its tip should also be grounded. Microelectronic devices should never be plugged into, or unplugged from, their sockets with the power on.

The leads of ICs are not flexible in the general sense. They can be formed to meet the requirements of a specific application, but should not be indiscriminately twisted or bent. It is better to use a lead-bending fixture designed for the purpose, but long-nosed pliers will do, provided you:

(a) Restrain the lead between the bending point and the plastic case to prevent the lead loosening in the case;

(b) If bending sideways, bend only the narrow part of the lead;

(c) Don't bend nearer than three millimeters from the case;

(d) Don't make bends too sharp (minimum radius 2 mm);

(e) Avoid repeated bending of leads;

(f) Use care to prevent damage to lead plating.

It is also important that the ends of bent leads be straight to assure proper insertion through the holes in the printed circuit board.

After you've finished the soldering, inspect the board for any solder bridging between conductors and any damaged circuits. If you've performed the previous steps carefully there should be none. Then cut the pigtails short, remove any excess flux and clean the board with some more of the resist ink solvent as before. As a final step, spray both sides of the board with a protective coating of **polyurethane resin**. This will prevent oxidation of the copper conductors.

The PCB may be mounted component side uppermost with four standoffs inserted through the four corner holes you drilled for the purpose, and four corresponding holes drilled in the cabinet base plate or chassis. The standoffs should raise the board about an inch clear of the base plate. However, when a board is provided with edge connectors, it is usually held in a cage or frame that keeps it locked in the external (female) connector.

Perfboards

The perfboard is a perforated phenolic board with holes drilled in it like a pegboard, but the holes are smaller and closer together so that the pins of DIP ICs can be inserted. This makes it easy to use for temporary or permanent circuits. Connections are made by wire wrapping.

Mount all items loosely by inserting their leads through the holes in the perfboard. It will be necessary to bend those of diodes and resistors. Do this carefully so that the bends are not too sharp and the leads are not loosened in any way. The components should be resting on the board with the leads going through perpendicularly.

This operation will be made easier if you have some kind of clamp to hold the board. (Radio Shack PC Board Holder, cat. no. 276-1568, is such an item.) This enables you to work on the board from both sides without having to hold it.

You also will need a wire-wrapping tool, wire stripper, and wrapping wire. Wire from AWG #20 to AWG #30 is used and comes with insulation in various colors. You don't need to have more than one color, however, as long as the circuit is not unduly complicated.

AWG #30 is popular for most circuits but it is easily broken, so don't get too ambitious with it. Also, its current carrying capacity is much less than that of ordinary hookup wire (AWG#20), so you may want to use the latter in a project where higher currents are involved. (The resistance of #30 is approximately 0.1 ohm per foot as compared to #20, which is approximately 0.01 ohm per foot.)

It is best at this point to trim off any excessively long leads if you can see that the residual length will still be adequate, but you must leave enough projecting from the board to accept from five to seven wraps of wire around the pin or lead. This requires the removal of about one inch of insulation from the end of the wire.

When you have completed the assembly of the components on the perfboard, and have verified that the circuit works correctly, you should solder the connections to the leads of resistors, transistors, capacitors, and so on, if the circuit is to be permanent. This is not essential for wires wrapped around the square-section pins of DIP sockets if you used them (which is a good idea, anyway), since they make a much tighter joint. (See Figure 10.2.)

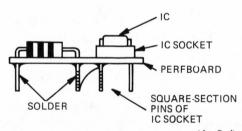

After Radio Shack Eng. Notebook

Figure 10.2 Mounting Components on a Perfboard.

Repairing a Printed Circuit Board

ICs are not themselves repairable and, if found to be malfunctioning, must be replaced. How this is done depends upon the original installation. Changing ICs mounted in sockets is almost as easy as replacing vacuum tubes. (A little extra care is necessary to avoid damaging the pins, which are not as sturdy as tube pins.) But those that are soldered in place on printed circuit boards can be quite a chore because of the number of pins that must all be unsoldered before they can be removed.

Desoldering an IC *can* be done with an ordinary pencil-type soldering iron and a "solder sucker." The latter is a syringe that sucks the liquid solder from the joint, leaving it free (in theory, anyway). The simplest type of solder sucker consists of a narrow tube fitted at one end with a rubber bulb. You apply the other end of the tube to the liquid solder to be removed, first squeezing the bulb. Then you release the pressure on the bulb so that the solder is sucked up the tube. A better type has a spring-loaded piston with a triggering mechanism. You trigger it when you want it to suck the solder, which is done by the piston's springing back up in its cylinder, creating a partial vacuum. But since ICs have so many pins, unsoldering them one by one is inefficient as well as tedious.

To overcome this, special soldering iron tips are available that contact all the pins at the same time, melting the solder on all of them simultaneously. However, you do have to be dexterous in using these, because the greater heat being applied may destroy the IC if you apply it for too long.

CHAPTER **11**

Troubleshooting
Digital IC Equipment

Why Troubleshooting ICs Is Different

There are fundamental differences between troubleshooting integrated circuit equipment and equipment employing discrete components. Discrete components have simple characteristics, such as resistance, capacitance and inductance, that are easily checked by traditional troubleshooting tools. But when the circuit is microminiaturized and encapsulated, these components are no longer accessible. The only thing you can test is the performance of the complete circuit.

The classical method of isolating the cause of a failure consists of four steps:

1. Determine symptoms
2. Determine section
3. Determine circuit
4. Determine component

Since we can't do Step 4, it seems our task has become simpler, and this is true to some extent of linear ICs whose performance can still be evaluated by use of signal generator and oscilloscope. But since the majority of ICs are digital, with numerous inputs and outputs to be stimulated and monitored simultaneously, Step 3 can become extremely time-consuming. If we have to disconnect leads and unsolder pins as well the resultant wear and tear is likely to have further adverse effects on boards and printed circuits, as well as increasing the chances of static burnout, as you saw in the previous chapter.

Linear circuits handle analog signals with voltage profiles corresponding to values that vary continuously, which is why the oscilloscope is the preferred instrument for studying them. Digital circuits, on the other hand, operate by switching between states—from high to low, and vice versa. To be in the high state is to have a potential above the high threshold; to be in the low state is to have a potential below the low threshold. The *exact* value of the potentials is unimportant as long as they are above or below the respective thresholds, since this is all that is needed to actuate the circuit. If a sufficient stimulus is present at the input, the circuit will produce the proper output every time if it is working properly.

From this it becomes clear that the essential requirement for troubleshooting equipment with digital ICs, is a means of verifying the presence of the correct voltage states on the various pins of the different ICs, without trying to remove them from the board, since this not only saves time and minimizes the risk of damage, but also allows us to examine the IC's performance in its natural environment.

As might be expected, the manufacturers of electronic test equipment have not been slow in coming up with devices designed for the purpose. First though, let's consider types of failures to look for, and symptoms produced by them.

Common Defects

IC failures fall into two main categories: external and internal. External failures occur in connections between ICs or in discrete components connected to them. These are repairable. Internal failures are not.

The commonest internal defect is an open bond between the pin and the chip. When this happens to an input bond (as in Figure 11.1), the correct signal will be present on the pin, but cannot reach the chip. The input of the chip will float to a "bad" level between the high and low thresholds, which a TTL gate will see as a permanent high state. The effect on the output will depend upon the nature of the circuit. If we take as an example a NAND gate with two

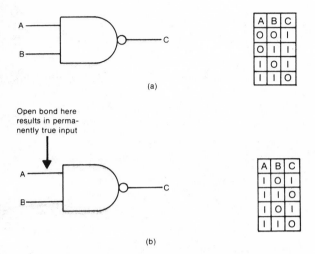

Figure 11.1 Effect of open bond on NAND gate input.

inputs with a normal truth table as at (a), the effect of an open bond at A will be the truth table at (b).

An open output bond will block the correct output from the chip so that it will not appear on the output pin. This will also go to a bad level, and will affect any and all input pins on other ICs to which it may be connected. These will behave as if they had open input leads, with the difference that the "bad" level will now be present on the input pin. This would lead you to look further back for the origin of the malfunction.

Instead of opening up, the input or output bond may short to V_{CC} or ground. If the short is to V_{CC}, all signal lines connected to that point will be high. Conversely, a short to ground will hold them to a low state. The most likely effect of either condition is to inhibit all signals normally to be found beyond this point, so it is one of the easiest troubles to track down.

It is not so easy to analyze a problem caused by a short between two pins when neither is V_{CC} or ground. When either pin goes to a low state, it pulls the other with it, yet when the two pins should be high or low together, they show the proper voltage.

An internal failure in the chip is always catastrophic to its performance, so that the output pins are locked high or low and will not change in response to appropriate stimuli. This is because the failure blocks the signal flow by completely preventing switching action.

An open signal path in the external circuit has the same effect upon the input to which it is connected as an open output bond in the output of the previous IC. The input will float to a "bad" level. However, as the correct signal appears on the output pin, we know that the interruption must exist between the output pin of the preceding IC and the input pin of the following IC.

A short between an external signal path and V_{CC} or ground exhibits the same symptoms as an internal short of the same kind. If the short is to V_{CC}, the signal path will be high at all times; if to ground, it will be permanently low. This one is hard to isolate, and only a very close inspection of the circuit will determine whether the fault is internal or external.

It is clear from the foregoing that troubleshooting equipment using ICs consists of the following three steps:

1. Observe closely the manner in which the behavior of the system differs from normal. This requires that you know what its normal behavior should be, which you would do if you were thoroughly familiar with it. If you are not, you must get hold of the service manual and study it. If this manual is any good (and service manuals are usually very good), it will also give the key signals to be found at various test points.

2. Using the information acquired so far, narrow the area of search to as few ICs as possible. Look for incorrect signals or voltage levels between ICs in an endeavor to isolate the trouble to a single IC.

3. Check the performance of the suspected IC or ICs.

The principle test devices used for this stage are:

1. Logic probe
2. Logic pulser
3. Logic clip
4. Logic comparator

Using a Logic Probe

The **logic probe** (Figure 11.2) has a fine tip that, when placed

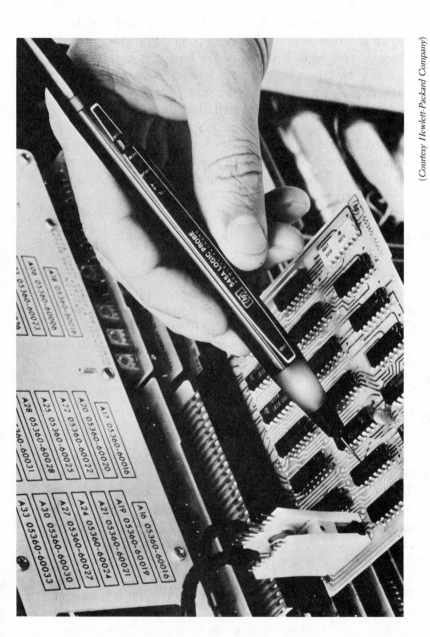

Figure 11.2 Logic probe that can be used for both TTL and MOS circuits, and that can memorize single pulses.

(Courtesy Hewlett-Packard Company)

in contact with one of the pins of an IC, or with a *node* (junction point) of the external circuit, allows you to see the digital state of that point. In the type illustrated, the lamp indicator glows brightly if the voltage level is high. If it is at a low level, the lamp does not light. At an in-between potential ("bad" level) it glows dimly. It also blinks to indicate the presence of a pulse train.

The logic probe has a power cord that can be connected to V_{CC} in the circuit being tested, or to a separate power supply. A probe designed for use with the voltage levels in TTL and similar circuits requires a 5-volt supply, while one for MOS circuits would need from 12 to 25 volts. Probes are available that can be changed from one level to another by means of a switch.

Using a Logic Pulser

The **logic pulser** resembles the probe in appearance. You apply the tip to a pin or node and press the pulse button to inject a narrow pulse into the circuit. When this pulse is applied to an input pin, it should cause the device to switch to its opposite state, or otherwise, as shown in the truth table. By connecting the logic pulser to an input (or inputs) and the logic probe to the appropriate output, you can test the operation of a logic gate very easily.

Using a Logic Clip

The **logic clip** (Figure 11.3) has a considerable advantage over the probe when the ICs in the circuit are the usual dual in-line package (DIP) with 14 or 16 pins, since it can display the state of all the pins at the same time. Constructed somewhat like a broad clothespin, it contacts all the IC pins at once. Its internal circuitry enables it to locate the V_{CC} supply and ground return automatically for its own use, and to display the condition of all the pins simultaneously by means of LEDs that light for each pin that is above a threshold level. The block diagram for the circuit of a logic clip is shown in Figure 11.4.

This is of great help where you must view the state of more than one pin at a time. For instance, a decade counter cannot be checked unless you can see what is happening on at least one input

(*Courtesy Hewlett-Packard Company*)

Figure 11.3   Logic clip (see text).

and four outputs. Used in conjunction with a logic pulser probe, the logic clip makes it possible to jump rapidly from point to point, applying stimuli and observing responses; a valuable capability in modern circuits employing many ICs, where most of the time you are troubleshooting you are looking at the good signals, trying to get *to* the problem.

Using a Logic Comparator

The tools we have discussed so far operate by checking outputs against inputs, which is standard troubleshooting procedure. The **log-**

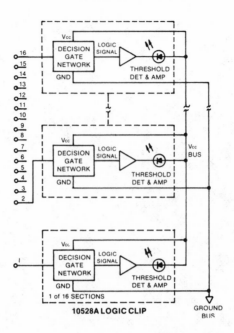

Figure 11.4   Circuit diagram of logic clip.

ic comparator, however, works by comparing the performance of the
IC in the circuit with an identical IC which is known to be good.
The reference IC may be mounted on a printed circuit board (or
"card") that is inserted in the comparator, or it may be plugged into
a socket on the side of the comparator, depending upon the way the
comparator is designed. The comparator, or in some cases, a clip on
the end of a lead connected to it, is then clipped on to the circuit
IC so as to contact all 14 or 16 pins. This provides that the reference IC gets the same inputs and power supply as the circuit IC.
This means that each pin on the reference IC should have exactly
the same signal on it as does the corresponding pin on the circuit
IC. If there is a discrepancy, the pin is indicated by a LED. If everything is correct, no light should appear.

One model of logic comparator also has a selector knob to connect any pin to a separate indicator, so that the state of that pin can
be ascertained, since it will not give any indication if the circuit and
reference ICs are the same. This logic comparator, therefore, in-

cludes the functions of a logic probe as well as its comparator function.

If you have a logic comparator you can save a lot of time by using it to test *all* testable ICs in the section where you suspect the trouble to be, noting which ICs and pins are discrepant. The word "testable" is used because not all models can test all ICs (the instruction manual will list them). Some ICs also require special conditions for testing, but fortunately there are very few of these and you don't come across them very often. The vast majority can be tested in their circuits without difficulty. However, if you do get a discrepancy, make sure that it is a real one and not caused by the clip not making good contact with the pins. Squeeze its jaws together and rock it slightly to make sure the contact is firm (some circuits have been sprayed with a sealer).

Determining Circuit in Trouble

This preliminary reconnaissance focuses attention on the problem area. The next step is to track down the origins of the discrepancies noted. The logic probe can be very useful here. With it you can check the signal activity on inputs and outputs to verify correct operation of the IC. For instance, if the IC is a DCU and the enabling inputs are in the enabled state, it should be counting. If the logic probe shows pulse activity on the outputs, you can assume it is working properly, since when a DCU fails it quits altogether.

The first failure to test for is an open bond in the IC driving the failed node. As we noted earlier, an open output bond causes the output pin to float to a "bad" level. If you find this to be the case, the IC must be replaced.

However, if the node is not at a bad level, you should look next for a short to V_{CC} or ground. The best way to do this is with a logic pulser and logic probe. The pulser does not have enough power to override either of these shorts, so if both probes are applied at the same time to the same node, no pulse will be indicated by the logic probe. If a pulse is indicated, there can be no short.

The short, if one exists, could be either internal or external. An external short would be caused by a solder bridge, loose wire or a defective discrete component connected to the node. If a thorough examination of the external circuit fails to reveal such a cause, then the problem might be in any one of the ICs connected to the node. Start by replacing the IC driving the node, then the others, one at a time, until the bad one is discovered.

If a pulse is detected when using the pulser and probe as described above, then you should check for a short between two nodes. This can be done by applying the pulser to the failing node and probing each of the other nodes in the area with the logic probe. If a short exists, the logic probe will detect the output of the pulser at a node which is not supposed to be connected to that being pulsed. (To check, reverse the positions of the logic probe and pulser and retest.)

Of course, another method of checking for a short is to unplug the board and use an ohmmeter. This would be the method if a pulser were not available. A bench power supply adjusted to the proper output (5 V) can also be used with a passive probe as a substitute for a pulser, although it is not so handy. A passive probe can also be connected to the V_{CC} bus, and then touched to the driving point.

If all the preceding steps have failed to isolate the fault, there are only two possibilities left: an open input bond or internal breakdown in the suspected IC. Replacing the IC should eliminate the problem.

However, it is essential to test the IC after installation. If the circuit now works, you have repaired it. But if the same symptoms are still there, then the IC you replaced was *not* the cause of the trouble. You will have to go back over the procedure to see if you can find out what it was that escaped notice. This is why all steps of the testing process must be done carefully, and the results noted and completely evaluated to avoid waste of time in backtracking.

There is still one other possibility to be considered. If, after a thorough troubleshooting session, including fruitless replacement of the IC or ICs, the malfunction persists, it may be due to an open signal path in the external circuit. The symptom for this is a bad logic level on one or more input pins, but a good level on the output driving them. Some models of comparator will not reveal this symptom, therefore you might miss it in your preliminary reconnaissance. A logic probe, however, will enable you to track backward from the affected input pin along the printed circuit signal path until you reach the point where the bad level is replaced by the good level (high, low or pulsing). At this point there must be an open, probably a hairline crack in the conductor, a dirty or bent pin, open feedthrough, broken lead or cold solder joint.

So much for detecting faults through their symptoms. There will come a time, sooner or later, when you will find all the nodes operating correctly, yet the circuit still malfunctions. Replacing ICs does no good either. This is what happens when you have a logic feedback loop fault. Here you will need both a logic comparator and a logic probe. If the feedback loop is "stuck," the probe will show where the activity stops. If the loop is operating, then the comparator shows where the IC is not obeying its truth table.

Special Problems in Troubleshooting Microprocessor Equipment

Microprocessors are now appearing in over half of all printed circuit boards in production or under development. Manually applying input stimuli and observing the output signals of such boards would be a time-consuming task. You'd need to know the inner workings of each of the devices on the board, keep track of the functions performed as they interact, and evaluate the vast quantity of data generated. As printed circuit boards may contain as many as a hundred ICs, with thousands of pins, the job could take weeks!

In some boards, test requirements can be simplified by removing the microprocessor, but this is only marginally acceptable as a stop-gap measure when you don't have the proper test equipment, and you can really only do it when the μP is installed in a socket. There is also the risk of damaging the μP by handling it. Furthermore, you still have to do some functional testing with the μP plugged in because there is no other way to investigate timing problems.

Timing problems cause incorrect synchronization between parallel lines, or *data skew*, in which the data sequences get out of step. Microprocessor-based systems commonly have a data flow 32 bits wide, so you can see timing is critical.

Bits are also mis-set by *glitches*, which are random pulses that have no business being there, but have sufficient amplitude to cross the threshold of a device and activate it. Glitches may be noise spikes or other unwanted interference, and are often of such brief duration as to be hard to detect, yet have the same effect as legitimate logic pulses.

To troubleshoot these more advanced systems, logic analyzers are used.

Logic State Analyzers

Logic state analyzers are large and expensive test instruments designed to display on a CRT the actual data being processed, "freezing" it to allow examination of the data words or word containing the anomaly. This type of equipment would be quite out of place in a service shop, because it is used for data processing equipment (which means computers and the things they interface with), but some readers who work in data processing will be using it.

Logic state analyzers consist basically of a special kind of oscilloscope, or of a special instrument or plug-in to be used with one. The data is captured in a memory and displayed on the screen as a number of traces showing the high and low levels of the data train, or as data tables of 1s and 0s. There are advantages in both methods. Displaying the actual pulses allows you to see what is really happening electrically (with the capability of expanding or amplifying any portion you want to examine more closely), but the numerical display is easier to read and correlate with a computer program or tape.

There are two ways of getting the segment of the data train you wish to study on to the screen. In one, the instrument counts off a predetermined number of "events," then stores the data that is passing through when the count stops. An alternative version of this stores the data after a preset time interval.

The other method requires the prior entry in a register provided for the purpose of a data pattern, or "trigger word." This is a replica of a word in the data, so that when the word in the data arrives, it acts as a trigger (by matching the stored trigger word), and the data passing through at that moment is captured.

Triggering can be set so that on the arrival of the trigger word,

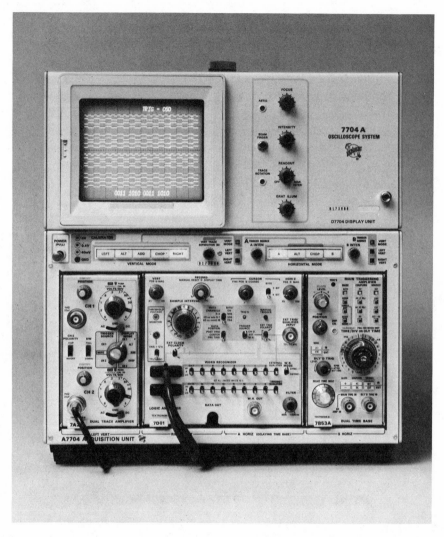

Figure 11.5 Oscilloscope with Logic Analyzer plug-in shows high and low levels of data train, and also (at bottom of screen) numerical equivalent of brightened portion. This brightened portion is selected by rotating the dial marked CURSOR, which moves it horizontally across the screen. The numbers 1 and 0 appear according to the state of each marked bit.

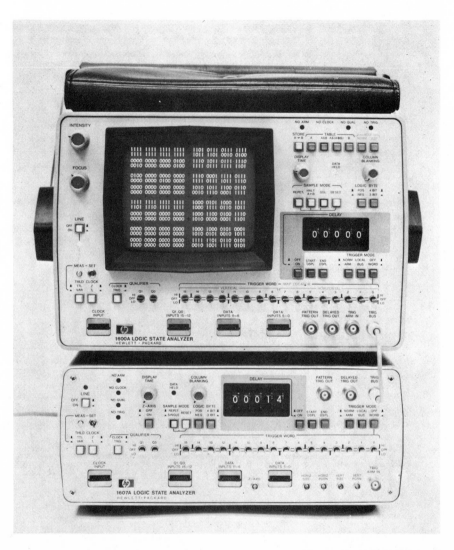

(*Courtesy Hewlett-Packard Company*)

Figure 11.6 Two Logic State Analyzers (one with a CRT) working together produce a 32-channel data domain display. The upper instrument can operate alone to
give a 16-channel display. The lower can be used with any modern scope to give a
16-channel display also.

it, and the next 15 words, are stored and displayed. Alternatively, the analyzer may be set so that the words preceding the trigger word may be displayed. This is particularly valuable, because it may show what led up to the error.

The two types of logic analyzers are illustrated in Figures 11.5 and 11.6.

CHAPTER 12

Troubleshooting Linear IC Equipment

General Troubleshooting Technique

Many technicians hesitate to tackle repair jobs on electronic equipment with which they are not familiar, feeling that the task may prove to be beyond their skills and knowledge. The experienced "old timer," on the other hand, may take on such jobs without hesitation, for he realizes that *all* electronic equipment of a given general type works on the same principles, can suffer the same defects, and is amenable to the same troubleshooting techniques. In fact, if we consider complaints in broad rather than limited terms, specific defects may cause the same complaints in all types of equipment. The physical manifestation of the "complaint" may vary considerably from one type of equipment to another, however.

For instance, an IC used as an amplifier can cause distortion due to a "leaky" transistor. This broad complaint (distortion) shows up in different ways according to what the amplifier is being used for:

(a) In the video section of a TV set, the distortion may show up as a *smeared picture.*

(b) In an audio amplifier or radio receiver, it may show up as *garbled sound.*

(c) In a radio transmitter, it may show up as *excessive harmonic radiation.*

(d) In a test instrument, it may show up as *inaccurate readings*.

(e) In an item of industrial control equipment, it may show up as *erratic operation*.

But in every case the defect is the same, the basic circuit is the same, the broad complaint is the same, and the defect can be found using the same general troubleshooting techniques (signal tracing, for example).

Thus, it is possible to provide the chart given in Table 12.1, which applies to all types of electronic equipment. It is used to select a suitable diagnostic test technique. To use this chart, first determine the *general complaint*. To do this you'll have to interpret the specific complaint in terms of its broad technical nature, as was done in the examples given above. For instance, if the equipment *doesn't work at all*, it is DEAD—whether the item is a receiver, audio amplifier, industrial control, transmitter or test instrument. If the equipment *works, but lacks its usual response*, it is WEAK. In a radio transmitter this may show up as lowered power output; in a receiver as lack of sensitivity; in control equipment as sluggish operation. If the equipment *works "now and then,"* or if some other complaint *"comes and goes,"* it is considered INTERMITTENT. If the equipment *works, but not quite properly,* the complaint is DISTOR-

DIAGNOSTIC TECHNIQUE / COMPLAINT	VISUAL INSPECTION	CHECK POWER SUPPLY	VOLTAGE CHECKS	SIGNAL TRACING	SIGNAL INJECTION	BRUTE FORCE	PARTS SUBSTITUTION	COMPONENT TESTS	ALIGNMENT	REMARKS
DEAD	X	X	X	X	X			X		
WEAK	X	X	X	X	X				X	Align only if other tests indicate
INTERMITTENT	X	X				X	X	X		
DISTORTION	X	X	X	X				X		"Weak" and "distorted" may be a common complaint
OSCILLATION	X	X		X			X		X	Similar tests for complaints of "hum" and "noise." Align only if other tests indicate

Table 12.1 Universal Troubleshooting Chart.

TION, as in the examples given above. If the equipment is *unstable*, the trouble is OSCILLATION. The test methods given for this complaint also apply where circuit operation is upset by an undesired signal such as *hum, noise* or *interference*.

In most cases any of several test techniques may be used, depending on the test instruments available and the type of equipment serviced. Two general checks should be made regardless of complaint: *visual inspection* and *power supply verification* (incorrect supply voltages can cause a variety of problems).

Where alignment is indicated by the table, this applies only to equipment with fixed tuned circuits.

Visual Inspection

The second table (12.2, on page 226) outlines in block diagram form, basic service procedure for all equipment. The first step, of course, is a quick visual inspection of the equipment, watching for obvious physical or electrical damage. This, in itself, may permit a quick isolation of the defect. At this time, check with the "customer" to see if the equipment has been subjected to unusual operating conditions, such as high temperature, humidity and so on.

Confirm Complaint

Always *confirm the complaint*. This means, try out the equipment to make sure the "complaint" is as described by the "customer." Too often you may be told "it doesn't work" when what is meant is that it doesn't work as well as it should. A technician would use this expression only to describe a dead set.

After the initial visual inspection and check of power supply (which includes batteries, of course), you'll select the appropriate static or dynamic tests to isolate the trouble to the particular stage and section.

Resistance Checks

Point-to-point *resistance checks* are useful for isolating defects which cause a change in DC resistance values. However, you need to

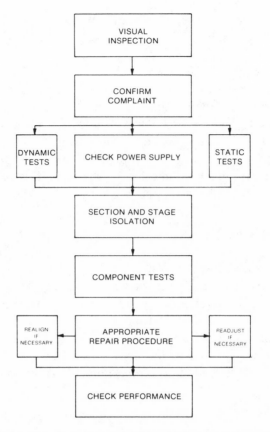

Table 12.2   Basic Steps in Servicing All Types of Equipment.

remember that in equipment employing transistors, misleading results can arise because of the direct resistive connection between a transistor's various electrodes. Since resistance tests apply only to the external circuit around the IC, use the following technique:

(a) Turn the equipment off.

(b) Discharge any large filter or bypass capacitors, using a short piece of wire or a clip lead.

(c) Remove the ICs if sockets are used; if soldered in place, remove them *only* from the stages to be checked.

(d) Using an ohmmeter, measure the DC resistance between each ICs connection points (each socket pin, for example, if sockets are used) and either circuit ground or common power connection.

(e) Compare the readings obtained with those given in the equipment's service manual or in the schematic diagram.

(f) Ignore *minor* differences between "actual" and "expected" values. *Major* differences indicate trouble.

DC Voltage Analysis

DC voltage analysis is perhaps the oldest of all test techniques, and is valuable for isolating defects that cause a change in the equipment's operating voltages. The procedure is as follows:

(a) Turn the equipment on, and adjust any controls for normal operation.

(b) Using a DC voltmeter, check voltages between each IC pin and "ground," being careful to observe proper polarity.

(c) Compare the readings obtained with the service manual or schematic.

(d) Ignore *minor* discrepancies. *Major* differences indicate trouble.

In addition to DC voltage tests, *current measurements* are sometimes helpful for locating troubles in precision control devices and instruments. An overall current test may be made as a check on equipment efficiency. For this, the meter is inserted in series with one of the power supply (or battery) leads.

Dynamic tests are tests made under actual operating conditions. These tests may be used to isolate all types of troubles, including not only serious component defects causing a major change in equipment operation, but minor defects which affect circuit efficiency and performance, but permit "nearly normal" operation. Dynamic tests are also valuable for tracking down intermittent defects.

Signal Tracing

Signal tracing is the most powerful technique, involving tracing a signal stage-by-stage as it passes through the equipment. The in-

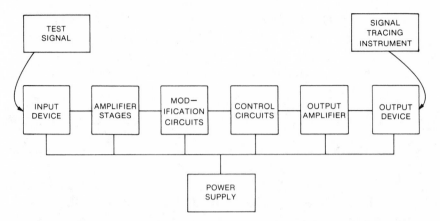

Figure 12.1 Signal Tracing Technique (see text). The test signal is applied where shown and the signal tracing instrument is connected to the input and output of each stage in turn. (This block diagram can be applied to almost any type of electronic equipment.)

strument used to follow the signal may be an AC voltmeter, a signal tracer or an oscilloscope. Generally, a signal tracer is used to follow signals through receivers and audio amplifiers, while an oscilloscope is superior for most other items. An oscilloscope can also be used to check receivers if it is provided with a demodulator probe or has the necessary bandwidth.

To use the signal tracing technique, as illustrated in Figure 12.1, proceed as follows:

(a) Turn the equipment on.

(b) Apply an adequate test signal to the equipment's input. In a receiver this signal may be obtained from an RF signal generator or by tuning in a local station. In an audio amplifier a suitable signal may be obtained from a record player, tuner or audio generator. In other cases, depending on the equipment being tested, the signal may be obtained from a pulse generator, square-wave generator, sawtooth oscillator or other device. If the equipment generates its own signal (for instance, a radio transmitter) a separate signal source may not be needed.

(c) Make sure that the connection of the signal source does not disturb the circuit. If necessary, insert a DC blocking capacitor, matching pad, or other device as necessary, in series with the input.

(d) Make sure the test signal does not overload the equipment. Always use the minimum signal needed for usable output.

(e) Use a signal tracing instrument (signal tracer, oscilloscope, etc.) to check the relative amplitude and quality of the signal at the input and output of each stage, as shown in Figure 12.1. Depending on the instrument used, the signal may be: (1) heard in a loudspeaker; (2) observed as the closure of a "tuning eye"; (3) indicated by the deflection of a meter pointer; or (4) seen as a waveform on a CRT.

(f) The test signal should be modified by each stage. In the case of an amplifier, for instance, the signal should be increased in amplitude. In a clipper stage, a portion of the signal should be removed or clipped off.

(g) If the signal is changed in an unexpected fashion (the amplitude drops instead of increasing, for instance), trouble is indicated and the defective stage has been isolated.

(h) Where necessary, change the type of pickup (probe) as the signal is followed through the equipment. For example, an RF detector probe is used for checking the RF and IF stages of a receiver, a direct probe for the audio section.

Waveform Analysis

Waveform analysis is used to analyze changes in equipment performance that are minor rather than major or catastrophic—a deterioration in frequency response, for instance. Here, the oscilloscope is the instrument to use. The test signal may be a sine wave, square wave, pulse, AM or FM carrier or other signal, depending on the nature of the test and the equipment being tested. To use this technique, apply a signal of accurately known characteristics and, using a 'scope, observe the signal waveform and the input and output of each stage. Changes in the signal waveform tell you a lot about stage operation. A "rounding" of a square wave, for example, indicates a falling off of an amplifier's high frequency response.

Signal Injection

Signal injection is another dynamic test technique. It is complementary to signal tracing, and is almost as powerful for tracking down trouble. As the name indicates, this technique involves injecting a test signal into the equipment:

(a) Connect an indicator to the equipment's output stage. This may be an AC voltmeter, oscilloscope, or even the equipment's own output device (loudspeaker, TV screen, etc.).

(b) Turn the equipment on.

(c) Apply an appropriate test signal to the input of the *output* stage. The test signal may be obtained from an RF signal generator when checking the IF and RF stages of radio receivers, or from an audio generator when checking the audio stages of receivers, P.A. amplifiers, hi-fi and similar equipment; or from a pulse or square-wave generator when testing for frequency response and so on.

(d) Make sure that the connection of the signal source does not change the normal operation of the stage by using a small coupling capacitor in series with the input lead if necessary.

(e) Adjust the input signal level to the minimum signal needed for a normal output indication; later, readjust this signal level as necessary to prevent overload. Do this by adjusting the signal generator's attenuator control.

(f) Transfer the signal injection point *back*, stage by stage, from the equipment's output stage to its input point (A in Figure 12.1).

(g) Change to a different type of test signal where necessary. For example, an audio signal for the audio stages, a modulated RF signal of the proper frequency for the IF stages and converter stage, and a modulated RF signal for the RF and antenna stages.

(h) Note changes in equipment's output as shown by the output indicator. Normally, output level will go *up* as the signal injection point is transferred back, requiring readjustment of the signal generator's attenuator.

(i) Unexpected changes in the equipment's output signal indicate that the defective stage has been isolated. In the case of
a "dead" stage, for example, the output signal will drop to
zero when the injection point is shifted from the output to
the input side of the stage.

Brute Force

Brute force testing means lightly tapping, wiggling, or otherwise manipulating the circuitry to make it exhibit an intermittent
symptom. Electronic equipment that has been misbehaving in this
manner usually is as good as gold as soon as it gets on the bench,
and must be stimulated into bad performance. Some types of intermittent behavior are temperature-related. In this case, you may get it
to act up by warming it in a box with a 40-watt light bulb as a heat-
er, or spraying suspected parts with freon to cool them below their
normal temperatures, depending upon what is required.

Part Substitution

Part substitution means, of course, replacing a suspected item
with another known to be good to see if that clears up the problem.
Obviously, if it does, you have isolated the defect. This is a hangover
from the old tube days, but still has a limited use in modern equipment.

Component Tests

Component tests of transistors, capacitors, resistors, transistors,
and so on may also be done using the appropriate test equipment.
This only applies to accessible items in the external circuit; obviously,
those items inside an IC cannot be tested separately!

Alignment

Alignment of television sets and FM receivers requires the use of a sweep generator and an oscilloscope. This technique is fully described in service manuals for the equipment.

Testing Operational Amplifiers

Most linear ICs are extremely simple to test, but operational amplifiers are somewhat less so on account of their feedback loops.

In the case of an operational amplifier with circuit as in Figure 12.2, with a closed-loop gain of X100, we would check the following parameters:

Rated voltage and current: Connect a resistor across the output approximating the designed load, and connect an oscilloscope across the resistor. Connect a signal generator and VTVM to the input and apply the proper V_{CC} to the appropriate pin.

Start with an input of 100 hertz at 0.1 mV and increase the amplitude until distortion appears on the oscilloscope display. Back off until you have a waveform with maximum amplitude without visible distortion, and measure its peak-to-peak value. This gives you V_{out}. The current is obtained from the formula:

$$I_{out} = \frac{V_{out}}{R_T}$$

R_T is the value of R_O and R_L in parallel $\left(= \dfrac{R_O R_L}{R_O + R_L}\right)$

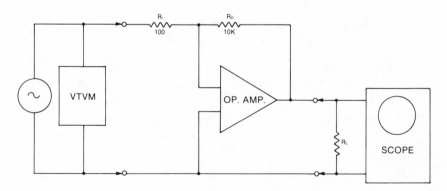

Figure 12.2   Testing an operational amplifier.

Frequency response: Readjust the signal generator for an output of 30 mV, then connect the VTVM to the output terminals instead of the oscilloscope. Increase the signal generator frequency until V_{out} has decreased by 3 dB (or to 70 percent) of original output voltage. This gives the maximum frequency response or bandwidth.

The remaining parameters are measured with the input grounded.

Input voltage offset: Measure V_{out} on the oscilloscope connected across the load resistor as before.

$$V_{in}\,(\text{offset}) = \frac{V_{out}}{100}$$

Input voltage drift versus supply drift: Adjust the V_{CC} voltage for $\dfrac{90 \times V_{cc}}{100}$ and measure the peak-to-peak output voltage on the oscilloscope. Then readjust V_{CC} to $\dfrac{100 \times V_{cc}}{100}$ and measure the output voltage as shown on the oscilloscope. The difference between the first and second reading is termed $\triangle^V$ offset.

$$\text{Drift} = \frac{\triangle^V \text{ offset}}{20}$$

Input noise: Measure rms value of output noise (connect the VTVM, set to read AC, parallel with the oscilloscope).

$$\text{input noise} = \frac{\text{output noise}}{100}$$

APPENDIX I

Glossary of Microelectronic Terms

A

ABSOLUTE ZERO: Temperature of zero thermal energy, corresponding to −273.16 degrees Celsius.

ACCEPTOR: Impurity element (such as boron, from Group III of the periodic table of the elements, which has three valence electrons) that increases the number of holes in a semiconductor, resulting in a p-type material in which holes are the majority carriers.

ACTIVE ELEMENT: An electronic circuit element which exhibits power gain or transistance, as in a transistor.

ACTIVE SUBSTRATE: A substrate which contributes to at least one active element.

ALLOY PROCESS: A fabrication technique in which a small part of the semiconductor material is melted together with the desired metal and allowed to recrystallize. The alloy developed is usually intended to form a p-n junction or an ohmic contact.

ALUMINUM: Silvery, lightweight, easily worked metal, used extensively in electronic hardware. Symbol, Al; atomic weight, 26.98; atomic number, 13.

ANALOG CIRCUIT: A circuit with voltages that vary continually.

ANTIMONY: Bright silvery metal that expands on solidifying. Used as an alloy with many other metals, especially in castings. Symbol, Sb; atomic weight, 121.75; atomic number, 51.

ARSENIC: Gray, nonmetallic element, from Group V of the periodic table of the elements, used as a donor impurity (see Donor). Symbol, As; atomic weight, 74.9216; atomic number, 33.

ASPECT RATIO: The ratio of the length of a microelectronic element divided by the width.

B

BIPOLAR TRANSISTOR: Junction transistor employing both types of charge carrier (electrons and holes).

BORON: Black, lustrous semiconductor, from Group III of the periodic table of the elements, used as an acceptor impurity (see Acceptor). Symbol, B; atomic weight, 10.811; atomic number, 5.

BREAKDOWN VOLTAGE: The value of voltage whose application to a p-n junction or dielectric causes voltage breakdown. When applied to a p-n junction, the breakdown voltage causes current flow to increase rapidly from a relatively low value to a relatively high value. When applied to a dielectric, the breakdown voltage causes the dielectric to act as a conductor rather than as an insulator.

BURIED-LAYER TECHNIQUE: A technique for fabricating monolithic transistors in which a low-resistivity material is introduced into the substrate under the transistor collector to reduce the effective collector resistance.

BYTE: Set of eight bits (binary digits).

C

CARRIERS: Electrons (n-type carriers) or holes (p-type carriers) in a semiconductor material which are available for the conduction of electric current.

CERAMIC: A product composed of inorganic, nonmetallic compounds formed through heat-processing. *Example*: alumina.

CHANNEL: An electrically-conductive path between two separated regions of a semiconductor material. In a field-effect transistor, it is the fabricated or induced path between the source and drain. In a bipolar transistor, the term is generally used to describe an undesirable localized, severely-inverted region of the surface.

CHANNEL DEPTH: The depth of a conducting channel, usually between the source and drain of a metal-oxide-semiconductor field-effect transistor.

CHIP: Tiny piece of semiconductor material scribed away from a wafer, upon which an electronic circuit, or circuits, has been formed.

CIRCUIT: The interconnected combination of a number of microelectronic elements or electronic parts to accomplish a desired function.

CONDUCTION BAND: Range of energies that freely moving electrons have within the structure of a solid.

CONDUCTIVE PATTERN: A network of conductive material distributed over the surface of a substrate for the purpose of interconnecting the circuit elements.

CONDUCTIVITY, N-TYPE: The conductivity associated with electrons in a semiconductor material.

CONDUCTIVITY, P-TYPE: The conductivity associated with hole carriers in a semiconductor material.

CONDUCTOR: A circuit element, such as metal or low-resistivity semiconductor material, whose function is to conduct electrical current.

CONTAMINANT: A foreign material present in a device which may change the device parameters. *Examples*: moisture and chemical ions.

COVALENT BONDING: Type of linkage between two atoms that arises from the electrostatic attraction of their nuclei for the same electrons. Also known as an electron-pair bond.

CRYSTAL LATTICE: Regular arrangement of atoms in a solid, determined by their chemical bonding.

CZOCHRALSKI CRYSTAL: A crystal grown by slowly withdrawing a seed crystal from a melt, while the melt is held slightly above the melting point of the material. High-quality, low-dislocation-density crystals of germanium and silicon are grown in this manner.

D

DEBOUNCER: Circuit that prevents spurious repetition by introducing a slight delay after the signal produced by depression of a keyboard key.

DENSITY, CARRIER: The number of hole and free-electron carriers per unit volume in a semiconductor material.

DEPLETION REGION: A region in a semiconductor material, usually at the surface, in which the majority carrier density is reduced from the bulk density.

DEPLETION REGION (JUNCTION): A space-charged region in which essentially neither holes nor electrons are available for conduction. *Example*: the reverse-biased region of a p-n junction.

DEPLETION-TYPE MOSFET: A field-effect transistor in which a shallow, electrically-conductive channel connecting the source and drain is diffused into the device during fabrication. This channel, of the same conductivity type as the source and drain, can be made more resistive by voltage-biasing the gate.

DEVICE: An electronic part consisting of one or more discrete active or passive elements.

DCTL (DIRECT-COUPLED TRANSISTOR LOGIC): A standard, common-emitter circuit in which all collectors are tied together and returned to the collector voltage supply through a common mode resistor.

DIE (DICE): See CHIP.

DIE BONDING: The attachment of a die to a header or to a device package.

DIELECTRIC: An insulating or nonconductive material used in microelectronic devices to fabricate capacitors and to insulate conductors.

DIELECTRIC ISOLATION: The achieving of electrical isolation of the monolithic integrated circuit elements from each other by dielectric material rather than by reverse-biased p-n junctions.

DIFFUSION PROCESS: The process of doping semiconductor materials by injecting an impurity into the crystal lattice at an elevated temperature. This process is usually performed by exposing the semiconductor crystal to a controlled surface concentration of dopants.

DIGITAL IC: An integrated circuit that performs a logic function.

DIODE: Junction between n and p regions in a semiconductor that passes majority-carrier current readily when biased in a forward direction (p region positive and n region negative), but can only pass minority carriers when reverse biased. Used as a rectifier or detector. Special diodes, such as zener diodes, varactors, and so on, are used for clamping, voltage regulation, frequency modulation, microwave signal generators, etc.

DIP SOLDERING: A process whereby a printed wiring board or substrate is brought into contact with the surface of molten solder by immersing the board for the purpose of simultaneously depositing solder onto the conductive surface or for soldering parts to the printed wiring.

DISCRETE ELEMENT: An individual component, such as a resistor or transistor, not part of an IC.

DISLOCATIONS, CRYSTAL: Imperfections in the idealized crystal structure.

DONOR: Impurity element (such as arsenic, from Group V of the periodic table of the elements) that provides (donates) an electron not required for crystal-lattice bonding. The electron may be free to conduct electric current.

DOPANT: See DONOR; and ACCEPTOR.

DOPANT CONCENTRATION: The acceptors per unit volume less the donors per unit volume in a p-type semiconductor material, or the donors per unit volume less the acceptors per unit volume in an n-type semiconductor material.

DRAIN: The section of a field-effect transistor in which the flow of charge carriers terminates. The drain corresponds to the collector of a bipolar transistor or the plate of a vacuum tube.

DRIFT MOBILITY: The average drift velocity of carriers per unit electric field in a homogeneous semiconductor.

DTL (DIODE TRANSISTOR LOGIC): A logic circuit using input coupling diodes to the base of a grounded-emitter circuit.

E

ECL (EMITTER-COUPLED LOGIC): A logic circuit in which all emitters share a common-emitter resistor.

ELECTRON-BEAM EVAPORATION: An evaporation technique in which the heating of the evaporant is achieved by electron bombardment.

ELECTRONVOLT (eV): Unit of energy of charged particles accelerated by electric fields, equal to 1.6×10^{-19} joule, and is the kinetic energy acquired by a particle bearing one unit of electric charge (1.6×10^{-19} coulomb) that has been accelerated through a potential difference of one volt.

ELECTROSTATIC FIELD: Region around an electric charge in which an electric force exists and which interacts with any other electric charge that may be present. The field is represented by force lines that start on positive charges and terminate on negative charges.

ENERGY GAP: The region between the valence and conduction bands in which no energy levels exist.

ENHANCEMENT-TYPE METAL-OXIDE-SEMICONDUCTOR FIELD-EFFECT TRANSISTOR: A common type of metal-oxide-semiconductor field-effect transistor. Its source-to-drain conduction is dependent upon a channel which is electrostatically induced by the gate potential.

EPITAXY, OR EPITAXIAL DEPOSITION: The growth of additional material on a substrate; usually a thin film having a crystal structure and orientation aligned with that of the substrate.

ETCHANT: A solution used by chemical reaction to remove the unwanted portion of a material bonded to a substrate or circuit board.

EVAPORANT: The material deposited on a substrate by evaporation.

EVAPORATIVE DEPOSITION: The technique of condensing a thin film of evaporated material upon a substrate. The source of the evaporant is a material usually heated in a high vacuum.

EXTRINSIC PROPERTIES (SEMICONDUCTOR): The resultant properties of a semiconductor material after modification by dopants, traps, dislocations, mechanical stress, and electrical fields.

F

FAN-IN: The number of similar type inputs connected into a microelectronic device. *Example*: Six logic inputs connected to one AND gate would have a fan-in of six.

FAN-OUT: The number of inputs (loads) which are driven by the output of a microelectronic device.

FIELD-EFFECT TRANSISTOR (FET): A semiconductor device in which the output current is controlled by an electrical field (See also JUNCTION FIELD-EFFECT TRANSISTOR and INSULATED-GATE FIELD-EFFECT TRANSISTOR.)

FLIP-CHIP: See REGISTRATIVE BONDING.

FLIP-FLOP: A circuit which has two stable states; commonly called a bistable multivibrator.

FOUR-LAYER DIODE (P-N-P-N): A diode consisting of four alternate layers of n-type and p-type material forming three p-n junctions which can exhibit silicon-controlled rectifier-like characteristics.

FREE ELECTRON: In this text, an electron that has been excited into the conduction band. This n-type carrier is not really at all like a free electron in space, for it is strongly influenced by the crystal lattice.

FRIT (GLASS): Finely ground glass, suspended in an organic vehicle and fired to form package parts such as headers.

FULL GATE: A MOSFET gate which extends the entire distance between and overlaps the source and drain regions.

G

GAIN-BANDWIDTH PRODUCT: The product of the closed-loop gain of an operational amplifier and its corresponding closed-loop bandwidth.

GALLIUM ARSENIDE (GaAs): Semiconductor material used for light-emitting diodes (LEDs).

GATE: The region of a field-effect transistor which controls the output current. The gate corresponds to the control grid of a vacuum tube and to the base of a bipolar transistor.

GERMANIUM: Silvery-gray metalloid semiconductor element, used for transistors, diodes, and so on. Symbol, Ge; atomic weight, 72.59; atomic number, 32.

H

HEADER: The portion of a device package to which the chip is attached and from which the external leads extend. *Examples*: TO-5 case header and flat-pack base.

HEAT SINK: A mechanical device used to absorb or transfer heat away from an element or part.

HOLE CONDUCTION: When a valence electron moves (under the influence of an electric field), it leaves a hole (vacancy) in its original position and fills another. The holes appear to flow in a direction opposite to the direction of the electron flow.

HYDROFLUORIC ACID (HF): Also known as hydrogen fluoride, a gaseous compound of hydrogen and fluorine (technically an acid only when dissolved in water). Capable of attacking glass, hence cannot be kept in glass bottles. Used for etching or frosting glass, and for dissolving silicon dioxide in the fabrication of ICs.

I

IMPURITY, ACCEPTOR: See Acceptor.

IMPURITY, DONOR: See Donor.

INDIUM: Brilliant, silvery-white metal in Group III of the periodic table of the elements, used as an acceptor impurity. Symbol, In; atomic weight, 114.82; atomic number, 49.

INJECTION, CARRIER: An increase in the concentration of minority carriers in a semiconductor material as a result of the application of voltage or radiation. *Example*: the introduction of minority carriers into the base region of a bipolar transistor from the emitter.

INSULATED-GATE FIELD-EFFECT TRANSISTOR (IGFET): A field-effect transistor in which the gate is insulated from the source-to-drain channel by a dielectric material. *Example*: MOSFET device with an oxide serving as the insulating dielectric.

INSULATOR: A material which essentially does not conduct current.

INTEGRATED CIRCUIT (IC): An electronic device containing integral active or passive elements, often in great numbers, which perform all or part of a circuit function.

INTRINSIC PROPERTIES (SEMICONDUCTOR): The properties of a semiconductor which are characteristic of the pure crystal.

INVERSION LAYER: A surface layer of a semiconductor material which has been inverted from its original conductivity type to the opposite conductivity type.

ISLAND: An isolated region diffused or alloyed into a semiconductor substrate.

ISOLATION DIFFUSION: A diffusion process which provides junctions to electrically isolate the various elements in the monolithic circuit.

J

JUNCTION FIELD-EFFECT TRANSISTOR: A common type of field-effect transistor in which the drain-to-source conduction is controlled by a p-n junction located between the source and the drain. The voltage applied to the p-n junction varies the depth of its depletion region which, in turn, controls the source-to-drain conduction.

JUNCTION, P-N (SEMICONDUCTOR): A region of transition between p- and n-type semiconductor material which exhibits an asymmetrical conduction, exploited in semiconductor devices.

L

LINEAR IC: An integrated circuit that performs an analog function.

LSI: Large-scale integration.

M

MAJORITY-CARRIER DEVICE: A semiconductor device which depends for its operation on the action of majority carriers. *Example*: MOSFET.

MAJORITY CARRIERS: Carriers of the same polarity as the host semiconductor material (holes in p-type material or electrons in n-type material).

MASK: An implement used to shield selected portions of a wafer or substrate during a deposition process; also, to shield selected portions of photosensitized material during photo processing.

METAL: As contrasted to semiconductor material, the conduction band of metal extends into the valence band and no energy gap exists. Metal conducts by free-electron movement and usually exhibits a positive resistivity-temperature coefficient.

METALLIZATION: The deposition of a metal film on a substrate by evaporative deposition, cathodic sputtering, vapor plating or electroplating.

METAL-OXIDE-SEMICONDUCTOR (MOS): A term employed with other nomenclature to describe the technique of insulating a metallized electrode from a semiconductor material, using an oxide dielectric.

METAL-OXIDE-SEMICONDUCTOR FIELD-EFFECT TRANSISTOR (MOSFET): A common type of insulated-gate field-effect transistor (IFGET) employing the metal-oxide-semiconductor (MOS) technique. The metallized gate is isolated from the drain-to-source channel by silicon dioxide.

MICROCOMPUTER: A microelectronic computer.

MICROELECTRONICS: That branch of electronics which is associated with extremely small electronic parts, assemblies, or systems.

MICROMINIATURIZATION: The utilization of the microelectronic art to fabricate microelectronic systems and functions.

MICROPROCESSOR: A computer's central processing unit formed on one chip.

MINORITY-CARRIER DEVICE: A semiconductor device which depends for its operation on the action of minority carriers.

MINORITY CARRIERS: Carriers of opposite polarity to the host semiconductor material (holes in n-type material, or electrons in p-type material).

MOBILITY (DRIFT AVERAGE): The measure of the average velocity of a carrier moving under the influence of an electric field.

MONOLITHIC CIRCUIT: A circuit with all elements fabricated on one substrate.

MOS CIRCUIT: Metal-oxide-semiconductor circuit. A circuit fabricated utilizing MOS devices. Such a circuit consists entirely or almost entirely of metal-oxide-semiconductor field-effect transistors, some of which are used as transistors, and others as resistors. MOS circuits are generally characterized by the large number of closely-packed field-effect transistors.

MSI: Medium-scale integration.

MULTIPLE-GATE DEVICE: A field-effect transistor (generally of the metal-oxide-semiconductor type) which has several gates. These gates may be arranged in series or in parallel between the source and drain of the transistor.

N

N-CHANNEL FIELD-EFFECT TRANSISTOR: A field-effect transistor in which the source, drain, and the conductive channel between them are of n-type conductivity material.

N-DIFFUSION: A semiconductor diffusion operation which results in an increase of the relative number of n-type carriers available. This is usually an indiffusion of donor impurities.

N-TYPE CARRIERS: Free electrons in a semiconductor material.

N-TYPE SEMICONDUCTOR: A semiconductor material in which conduction electrons are present in excess of holes.

O

OHMIC CONTACT: A contact between two materials in which the potential difference across them is proportional to the current passing through, and in which the voltage drop is the same for current flow in conduction either direction.

OHMS PER SQUARE: The unit of sheet resistivity. See SHEET RESISTIVITY.

OXIDATION (CHEMICAL): A chemical reaction resulting in an increase in oxygen, or in an acid-forming element or radical in a given compound. In microelectronics, silicon dioxide passivations are sometimes formed by chemical oxidation of the silicon surface.

OXIDATION (THERMAL): Chemical oxidation accelerated by the application of heat. Silicon dioxide layers of monolithic integrated circuits are usually formed by thermal oxidation.

OXIDE STEPS: A sharp variation in thickness of the silicon dioxide on the surface of an integrated circuit due to the selective removal of the oxide at various fabrication steps.

P

PADS: Metallized areas on the surface of a wafer or chip to which bonds, interconnects, or test probes may be applied.

PARASITIC ELEMENT: An extraneous or undesirable circuit element existing in a microelectronic circuit as a result of the inherent nature of the particular fabrication process and circuit element layout involved. *Examples*: stray capacitances or substrate transistor actions.

PARASITIC SHUNT CAPACITANCE: An undesirable parasitic capacitance in a monolithic integrated circuit often resulting from a reverse-biased isolation diode ("tub").

PART: An item that cannot normally be disassembled without destruction.

PASSIVE ELEMENT: An electronic circuit element which exhibits neither power gain nor transistance. *Examples*: capacitors, resistors, and transformers.

PASSIVE SUBSTRATE: A substrate which does not contribute to an active electrical element.

PATTERN DEFINITION: The accuracy, relative to the original artwork, of the reproduction of pattern edges in integrated circuit elements.

P-CHANNEL FIELD-EFFECT TRANSISTOR: A field-effect transistor in which the source, drain, and conducting channel between them are of p-type conductivity material.

P-DIFFUSION: A semiconductor diffusion operation which results in an increase of the relative number of p-type carriers available. This is usually an in-diffusion of acceptor impurities.

PHOSPHORUS: Nonmetallic element that glows in the dark, from Group V of the periodic table of the elements, used as a donor impurity. Symbol, P; atomic weight, 30.9738; atomic number, 15.

PHOTOMASK: See MASK.

PHOTO RESIST: A material which is photographically fixed in position to selectively mask against etching or plating. Typical photo resists include Kodak Photo Resist (KPR), Kodak Metal-Etch Resist (KMER), and Kodak Thin-Film Resist (KTFR).

PINCH-OFF VOLTAGE: The gate voltage of a field-effect transistor that blocks current flow in the depletion mode.

PLANAR DEVICE: A type of semiconductor device in which all p-n junctions terminate in the same geometric plane.

POLYCRYSTALLINE MATERIAL: A material, typically silicon or germanium, made up of many single crystals oriented randomly.

PRINTED CIRCUIT: A pattern comprising printed wiring formed in a predetermined design, formed on the surface of a substrate or board.

PRINTED CIRCUIT BOARD: A nonconductive board with a metallized conductor pattern and discrete components attached, to perform a circuit function.

PRINTED CONTACT: The portion of a printed circuit used to connect the circuit to a plug-in receptacle and to perform the function of a male pin of a connector.

P-TYPE CARRIERS: See HOLE CONDUCTION.

P-TYPE SEMICONDUCTOR: A semiconductor material in which holes are the majority carriers.

R

REGISTER MARK: An alignment mark used to establish the relative positions of superimposed masks during device fabrication.

RESIST: A material used to selectively mask against etching or plating.

RESISTIVE PATTERN: An array of resistive elements on a substrate.

RESISTIVITY: The electrical resistance across the opposite faces of a cube of material.

REGISTRATIVE BONDING: A method of attaching a component die to a substrate by inverting the die and bonding the die contacts to the thin-film pads of the substrate. The term "flip-chip" is sometimes used synonymously with registrative bonding.

RTL (RESISTOR-TRANSISTOR LOGIC): A logic circuit using resistors for input coupling to the base of a grounded-emitter transistor amplifier.

S

SEMICONDUCTOR MATERIAL: A material in which conductivity lies in the range between that of a conductor and an insulator. Semiconductor material has a negative resistance temperature coefficient over a given temperature range. The electrical characteristics of semiconductor materials (e.g., silicon, gallium arsenide, germanium) are dependent upon the small amounts of impurities (dopants) added.

SHEET RESISTIVITY: The electrical resistance measured across the opposite sides of a square of deposited thin-film material. Expressed in ohms per square.

SHELL: Electrons in an atom are thought of as occupying orbitals in the space surrounding the nucleus. The orbitals are regarded as distributed in diffuse "shells," the first (the K shell) being closest to the nucleus, followed by L, M, and so forth. The number of electrons that can occupy the first seven shells are, in sequence, 2, 8, 18, 32, 50, 72, and 98. In many cases, the outer shells are only partially occupied, and these determine the chemical properties of the atom.

SILICON: Hard, dark-gray semiconductor with a metallic luster, used for transistors, diodes, and so on. Symbol, Si; atomic weight, 28.086; atomic number, 14.

SILICON DIOXIDE: A dielectric material commonly used in the surface passivation of microelectronic circuits.

SINGLE CRYSTAL: A monolithic material in which the crystallographic orientation of all the basic groups of atoms is the same.

SLICE: See WAFER.

SOURCE: The region of a field-effect transistor in which the flow of majority carriers originates. The source corresponds to the emitter of a bipolar transistor or the cathode of a vacuum tube.

SPUTTERING: The ejection of atoms or molecules from the surface of a material, resulting from bombardment by atoms or ions, and utilized as a source of material for thin-film deposition.

STATES, SURFACE: Discontinuities and contaminants at the surface of a semiconductor device which tend to change the surface resistivity or the carrier mobility and lifetime. Surface states may cause inversion layers, accumulation layers, or device parameter instability.

SUBSTRATE: A material upon which epitaxial layers, thick-film depositions, or thin-film deposition are made, or within which diffusions are made.

SURFACE MOBILITY: Carrier mobility at the semiconductor surface. See MOBILITY, DRIFT.

T

THERMAL ENERGY: Random kinetic energy defined as the equivalent of heat. Absence of thermal energy coincides with absolute zero of temperature, and thermal energy increases with increasing temperature.

THERMOCOMPRESSION BONDING: The joining of materials by the combined effects of temperature and pressure.

THRESHOLD VOLTAGE: The voltage which must be applied to the gate of a metal-oxide-semiconductor field-effect transistor to create a source-to-drain conduction path (channel).

TRANSCONDUCTANCE (FET): The ratio of the variation of the source-to-drain current to the corresponding variation of the gate-to-source voltage, with the output voltage held constant.

TRANSISTANCE: An electronic characteristic exhibited in the form of voltage or current gain, or in the ability to control voltages or currents in a precise manner.

TTL (TRANSISTOR-TRANSISTOR LOGIC): A logic cicuit having all inputs connected to the multiple emitters of a single, common-base-connected transistor. The associated output transistor is used as an inverter amplifier.

TUB: The region of a monolithic integrated circuit providing electrical isolation of the circuit elements by means of the high impedance of a reverse-biased p-n junction or dielectric material.

U

ULTRASONIC ATTACHMENT: Die and lead attachments which utilize ultrasonic vibrations during the attachment operation.

UNIPOLAR TRANSISTOR: Transistor in which conduction is by carriers of one polarity only, as in a field-effect transistor.

V

VALENCE ELECTRONS: Electrons in the outermost shell of an atom that enter into the formation of chemical bonds. Being in the outer shell, they are more weakly attached to the nucleus, and thus can be shared, or excited into the conduction band.

VAPOR EPITAXY: See EPITAXY.

VOLTAGE BREAKDOWN: Rapid increase of current flow, from a relatively low value to a relatively high value, upon application of breakdown voltage to a p-n junction or dielectric.

VOLTAGE BREAKDOWN, AVALANCHE: A voltage breakdown that is due to the multiplication of carriers resulting from an accelerating field in a reverse-biased junction.

VOLTAGE BREAKDOWN (SECONDARY): A sudden reduction in the p-n junction breakdown voltage due to localized heating or imperfections in the device structure.

VOLTAGE BREAKDOWN (ZENER): A breakdown caused in a semiconductor device by the field emission of charge carriers in the depletion layer.

W

WAFER: A piece of substrate or semiconductor material that is wafer-like. *Example*: a thin silicon slice whose large surface areas are essentially parallel. See DIE.

WAVE SOLDERING: A soldering process whereby a mechanically induced wave of molten solder is brought into contact with the conductive surface of a printed wiring board or a substrate. The process also provides for the soldering of parts to the printed wiring board or substrate.

WEDGE BOND: A type of lead bond which is bonded with a wedge-shaped tool. A wedge bond may be a cold-weld, an ultrasonic, or a thermocompression bond.

Conversion Factors (U.S. and Metric)

Ampere-hour	3.6×10^{-3}	coulombs
Ampere-turn	1.257	gilberts
Angstrom	1.00×10^{-10}	meter
British thermal unit (BTU) (mean)	1.05587×10^3	joules
Centimeter	3.937×10^{-1}	inches
Coulomb	2.778×10^{-4}	ampere-hours
Foot	3.048×10^{-1}	meters
Gauss	1.00×10^{-4}	tesla
Gilbert	7.958×10^{-1}	ampere-turns
Grain	6.479891×10^{-5}	kilograms
Gram	3.527×10^{-2}	ounces (avoirdupois)
Horsepower (electric)	7.46×10^2	watts
Inch	2.54	centimeters
Kilogram	2.205	pounds (avoirdupois)
Kilometer	6.2137×10^{-1}	miles
Kilowatt-hour	3.6×10^6	joules
Liter	2.113	pints (U.S.)
Maxwell	1.00×10^{-8}	weber
Meter	1.094	yards
Micrometer	39.4	microinches*
Micron	1.00×10^{-6}	meter
Mil**	25.4	micrometers

---

* A microinch is a millionth of an inch.
** A mil is a thousandth of an inch.

Mile (nautical)	1.852	kilometers
Mile (statute)	1.609344	kilometers
Nanometer	3.94×10^{-2}	microinch*
Oersted	7.9577472×10^{-1}	amperes per meter
Ounce (avoirdupois)	28.349523125	grams
Ounce (troy)	31.1034768	grams
Pint (U.S.)	$4.73176473 \times 10^{-1}$	liter
Pound (avoirdupois)	4.5359237×10^{-1}	kilogram
Tesla	1.00×10^{4}	gauss
Watt	1.341×10^{-3}	horsepower
Weber	1.00×10^{8}	maxwells
Yard	9.144×10^{-1}	meter

---

* A microinch is a millionth of an inch.

Electronics Formulas
and Mathematical Tables

The following equations are those required most often in the field of electronics. The meanings of the symbols used are listed below (numerical subscripts are added when the same symbol is used for more than one quantity in the same formula).

A = length of side adjacent to θ in the right triangle, in same units as other sides
B = susceptance, in siemens
C = capacitance, in farads
D = dissipation factor
d = thickness of dielectric (spacing of plates), in centimeters
dB = decibels
E = potential, in volts
F = temperature, in degrees Fahrenheit
f = frequency, in hertz
G = conductance, in siemens
H = length of hypotenuse (side opposite right angle) in right triangle, in same units as other sides
I = current, in amperes
K = dielectric constant; coupling coefficient; or temperature, in kelvins
L = self inductance, in henries
M = mutual inductance, in henries
N = number of plates; or number of turns

O = length of side opposite to θ in right triangle, in same units as other sides

P = power, in watts

p.f. = power factor

Q = figure of merit; or quantity of electricity stored, in coulombs

R = resistance, in ohms

S = area of one plate of capacitor, in square centimeters

X = reactance, in ohms

X_c = capacitive reactance, in ohms

X_L = inductive reactance, in ohms

Y = admittance, in siemens

Z = impedance

δ = $90 - \theta$ degrees

θ = phase angle, in degrees; angle, in degrees, in right triangle, whose sine, cosine, tangent, etc., is required

λ = wavelength, in meters

π = $3.1416\ldots$

Admittance:

(1) $Y = \dfrac{1}{\sqrt{R^2 + x^2}}$

(2) $Y = \dfrac{1}{Z}$

(3) $Y = \sqrt{G^2 + B^2}$

Average value:

(1) Average value $= 0.637 \times$ peak value

(2) Average value $= 0.900 \times$ r.m.s. value

Capacitance:

(1) Capacitors in parallel:
$C = C_1 + C_2 + C_3 \ldots$ etc.

(2) Capacitors in series:

$$C = \dfrac{1}{\dfrac{1}{C_1} + \dfrac{1}{C_2} + \dfrac{1}{C_3} \ldots \text{ etc.}}$$

(3) Two capacitors in series:

$$C = \frac{C_1 C_2}{C_1 + C_2}$$

(4) Capacitance of capacitor:

$$C = 0.0885 \, \frac{KS(N-1)}{d}$$

(5) Quantity of electricity stored:

$$Q = CE$$

Conductance:

(1) $G = \dfrac{1}{R}$

(2) $G = \dfrac{I}{E}$

(3) $G_{total} = G_1 + G_2 + G_3 \ldots$ (resistors in parallel)

(4) $I_{total} = EG_{total}$

(5) $I_2 = \dfrac{I_{total} G_2}{G_1 + G_2 + G_3 \ldots \text{ etc.}}$

(current in R_2)

Cosecant:

(1) $\csc \theta = \dfrac{H}{O}$

(2) $\csc \theta = \sec (90 - \theta)$

(3) $\csc \theta = \dfrac{1}{\sin \theta}$

Cosine:

(1) $\cos \theta = \dfrac{A}{H}$

(2) $\cos \theta = \sin (90 - \theta)$

(3) $\cos \theta = \dfrac{1}{\sec \theta}$

Cotangent:

(1) $\cot \theta = \dfrac{A}{O}$

(2) $\cot \theta = \tan (90 - \theta)$

(3) $\cot \theta = \dfrac{1}{\tan \theta}$

Decibel:

(1) $dB = 10 \log \dfrac{P_1}{P_2}$

(2) $dB = 20 \log \dfrac{E_1}{E_2}$ (source and load impedance equal)

(3) $dB = 20 \log \dfrac{I_1}{I_2}$ (source and load impedance equal)

(4) $dB = 20 \log \dfrac{E_1 \sqrt{Z_2}}{E_2 \sqrt{Z_1}}$ (source and load impedances unequal)

(5) $dB = 20 \log \dfrac{I_1 \sqrt{Z_1}}{I_2 \sqrt{Z_2}}$ (source and load impedances unequal)

Frequency:

(1) $f = \dfrac{3 \times 10^8}{\lambda}$

(2) $f = \dfrac{1}{2\pi\sqrt{LC}}$

Impedance:

(1) $Z = \sqrt{R^2 + X^2}$

(2) $Z = \sqrt{G^2 + B^2}$

(3) $Z = \dfrac{R}{\cos \theta}$

(4) $Z = \dfrac{X}{\sin \theta}$

(5) $Z = \dfrac{E}{I}$

(6) $Z = \dfrac{P}{I^2 \cos \theta}$

(7) $Z = \dfrac{E^2 \cos \theta}{P}$

Inductance:

(1) Inductors in series: $L = L_1 + L_2 + L_3 \ldots$ etc.

(2) Inductors in parallel:

$$L = \cfrac{1}{\dfrac{1}{L_1} + \dfrac{1}{L_2} + \dfrac{1}{L_3} \ldots \text{etc.}}$$

(3) Two inductors in parallel: $L = \dfrac{L_1 L_2}{L_1 + L_2}$

(4) Coupled inductances in series with fields aiding:

$$L = L_1 + L_2 + 2M$$

(5) Coupled inductances in series with fields opposing:

$$L = L_1 + L_2 - 2M$$

(6) Coupled inductances in parallel with fields aiding:

$$L = \cfrac{1}{\dfrac{1}{L_1 + M} + \dfrac{1}{L_2 + M}}$$

(7) Coupled inductances in parallel with fields opposing:

$$L = \cfrac{1}{\dfrac{1}{L_1 - M} + \dfrac{1}{L_2 - M}}$$

(8) Mutual induction of two r-f coils with fields interacting:

$$M = \frac{L_1 - L_2}{4}$$

where L_1—total inductance of both coils with fields aiding

L_2—total inductance of both coils with fields opposing

(9) Coupling coefficient of two r-f coils inductively coupled so as to give transformer action:

$$K = \frac{M}{\sqrt{L_1 L_2}}$$

Meter
formulas:

(1)   Ohms per volt $= \dfrac{1}{I}$ (I = full-scale current in amperes)

(2)   Meter resistance: $R_{meter} = R_{rheostat}$

(The meter is connected in series with a
battery and a rheostat, and the rheostat
is adjusted until the meter reads full
scale. A second rheostat is then connected in parallel with the meter and adjusted until the meter reads half scale. The
resistance of the second rheostat will
equal that of the meter.)

(3)   Current shunt: $R = \dfrac{R_{meter}}{N - 1}$

where N is the new full-scale reading divided by the original full-scale reading
(both in the same units).

(4)   Voltage multiplier:

$$R = \frac{\text{Full-scale reading required}}{\text{Full-scale current of meter}} - R_{meter}$$

where reading is in volts and current in
amperes.

Ohm's law
formulas for
DC circuits

(1) $I = \dfrac{E}{R}$

(2) $I = \sqrt{\dfrac{P}{R}}$

(3) $I = \dfrac{P}{E}$

(4) $R = \dfrac{E}{I}$

(5) $R = \dfrac{P}{I^2}$

(6) $R = \dfrac{E^2}{P}$

(7) $E = IR$

$$(8) \quad E = \frac{P}{I}$$

$$(9) \quad E = \sqrt{PR}$$

$$(10) \quad P = I^2R$$

$$(11) \quad P = EI$$

$$(12) \quad P = \frac{E^2}{R}$$

Ohm's law formulas for AC circuits

$$(1) \quad I = \frac{E}{Z}$$

$$(2) \quad I = \sqrt{\frac{P}{Z\cos\theta}}$$

$$(3) \quad I = \frac{P}{E\cos\theta}$$

$$(4) \quad Z = \frac{E}{I}$$

$$(5) \quad Z = \frac{P}{I^2\cos\theta}$$

$$(6) \quad Z = \frac{E^2\cos\theta}{P}$$

$$(7) \quad E = IZ$$

$$(8) \quad E = \frac{P}{I\cos\theta}$$

$$(9) \quad E = \sqrt{\frac{PZ}{\cos\theta}}$$

$$(10) \quad P = I^2Z\cos\theta$$

$$(11) \quad P = IE\cos\theta$$

$$(12) \quad P = \frac{E^2\cos\theta}{Z}$$

Peak value:

(1) Peak value $= 1.414 \times$ r.m.s. value

(2) Peak value $= 1.570 \times$ average value

Peak-to-peak value:

(1) P-P value $= 2.828 \times$ r.m.s. value

(2) P-P value $= 3.140 \times$ average value

Phase angle: $\quad \theta = \text{arc tan } \dfrac{X}{R}$

Power factor:
(1) p.f. $= \cos \theta$
(2) $D = \cot \theta$

Q (figure of merit)
(1) $Q = \tan \theta$
(2) $Q = \dfrac{X}{R}$

Reactance:
(1) $X_L = 2\pi fL$
(2) $X_C = \dfrac{1}{2\pi fC}$

Resistance:
(1) Resistors in series: $R = R_1 + R_2 + R_3 \ldots$ etc.
(2) Resistors in parallel:
$$R = \dfrac{1}{\dfrac{1}{R_1} + \dfrac{1}{R_2} + \dfrac{1}{R_3} \ldots \text{ etc.}}$$
(3) Two resistors in parallel: $R = \dfrac{R_1 R_2}{R_1 + R_2}$

Resonance:
(1) $f = \dfrac{1}{2\pi\sqrt{LC}}$
(2) $L = \dfrac{1}{4\pi^2 f^2 C}$
(3) $C = \dfrac{1}{4\pi^2 f^2 L}$

Right triangle:
(1) $\sin \theta = \dfrac{O}{H}$
(2) $\cos \theta = \dfrac{A}{H}$
(3) $\tan \theta = \dfrac{O}{A}$

$$(4) \quad \csc \theta = \frac{H}{O}$$

$$(5) \quad \sec \theta = \frac{H}{A}$$

$$(6) \quad \cot \theta = \frac{A}{O}$$

Root-mean-square value

(1) R.m.s. value $= 0.707 \times$ peak value
(2) R.m.s. value $= 1.111 \times$ average value

Secant:

$$(1) \quad \sec \theta = \frac{H}{A}$$

$$(2) \quad \sec \theta = \csc (90 - \theta)$$

$$(3) \quad \sec \theta = \frac{1}{\cos \theta}$$

Sine:

$$(1) \quad \sin \theta = \frac{O}{H}$$

$$(2) \quad \sin \theta = \cos (90 - \theta)$$

$$(3) \quad \sin \theta = \frac{1}{\csc \theta}$$

Susceptance:

$$(1) \quad B = \frac{X}{R^2 + X^2}$$

$$(2) \quad B = \frac{1}{X}$$

$$(3) \quad B = B_1 + B_2 + B_3 \ldots \text{ etc.}$$

Tangent:

$$(1) \quad \tan \theta = \frac{O}{A}$$

$$(2) \quad \tan \theta = \cot (90 - \theta)$$

$$(3) \quad \tan \theta = \frac{1}{\cot \theta}$$

Temperature:

(1) $C = 0.556F - 17.8$
(2) $F = 1.8C + 32$
(3) $K = C + 273$

Transformer
ratio:

$$\frac{N_p}{N_s} = \frac{E_p}{E_s} = \frac{I_s}{I_p} = \sqrt{\frac{Z_p}{Z_s}}$$

(subscript p = primary; subscript s = secondary)

Wavelength:

$$\lambda = \frac{3 \times 10^8}{f}$$

Common Logarithms

N	0	1	2	3	4	5	6	7	8	9	N
10	0000	0043	0086	0128	0170	0212	0253	0294	0334	0374	10
11	0414	0453	0492	0531	0569	0607	0645	0682	0719	0755	11
12	0792	0828	0864	0899	0934	0969	1004	1038	1072	1106	12
13	1139	1173	1206	1239	1271	1303	1335	1367	1399	1430	13
14	1461	1492	1523	1553	1584	1614	1644	1673	1703	1732	14
15	1761	1790	1818	1847	1875	1903	1931	1959	1987	2014	15
16	2041	2068	2095	2122	2148	2175	2201	2227	2253	2279	16
17	2304	2330	2355	2380	2405	2430	2455	2480	2504	2529	17
18	2553	2577	2601	2625	2648	2672	2695	2718	2742	2765	18
19	2788	2810	2833	2856	2878	2900	2923	2945	2967	2989	19
20	3010	3032	3054	3075	3096	3118	3139	3160	3181	3201	20
21	3222	3243	3263	3284	3304	3324	3345	3365	3385	3404	21
22	3424	3444	3464	3483	3502	3522	3541	3560	3579	3598	22
23	3617	3636	3655	3674	3692	3711	3729	3747	3766	3784	23
24	3802	3820	3838	3856	3874	3892	3909	3927	3945	3962	24
25	3979	3997	4014	4031	4048	4065	4082	4099	4116	4133	25
26	4150	4166	4183	4200	4216	4232	4249	4265	4281	4298	26
27	4314	4330	4346	4362	4378	4393	4409	4425	4440	4456	27
28	4472	4487	4502	4518	4533	4548	4564	4579	4594	4609	28
29	4624	4639	4654	4669	4683	4698	4713	4728	4742	4757	29
30	4771	4786	4800	4814	4829	4843	4857	4871	4886	4900	30
31	4914	4928	4942	4955	4969	4983	4997	5011	5024	5038	31
32	5051	5065	5079	5092	5105	5119	5132	5145	5159	5172	32
33	5185	5198	5211	5224	5237	5250	5263	5276	5289	5302	33
34	5315	5328	5340	5353	5366	5378	5391	5403	5416	5428	34
35	5441	5453	5465	5478	5490	5502	5514	5527	5539	5551	35
36	5563	5575	5587	5599	5611	5623	5635	5647	5658	5670	36
37	5682	5694	5705	5717	5729	5740	5752	5763	5775	5786	37
38	5798	5809	5821	5832	5843	5855	5866	5877	5888	5899	38
39	5911	5922	5933	5944	5955	5966	5977	5988	5999	6010	39
40	6021	6031	6042	6053	6064	6075	6085	6096	6107	6117	40
41	6128	6138	6149	6160	6170	6180	6191	6201	6212	6222	41
42	6232	6243	6253	6263	6274	6284	6294	6304	6314	6325	42
43	6335	6345	6355	6365	6375	6385	6395	6405	6415	6425	43
44	6435	6444	6454	6464	6474	6484	6493	6503	6513	6522	44
45	6532	6542	6551	6561	6571	6580	6590	6599	6609	6618	45
N	0	1	2	3	4	5	6	7	8	9	N

Common Logarithms (Continued)

N	0	1	2	3	4	5	6	7	8	9	N
46	6628	6637	6646	6656	6665	6675	6684	6693	6702	6712	46
47	6721	6730	6739	6749	6758	6767	6776	6785	6794	6803	47
48	6812	6821	6830	6839	6848	6857	6866	6875	6884	6893	48
49	6902	6911	6920	6928	6937	6946	6955	6964	6972	6981	49
50	6990	6998	7007	7016	7024	7033	7042	7050	7059	7067	50
51	7076	7084	7093	7101	7110	7118	7126	7135	7143	7152	51
52	7160	7168	7177	7185	7193	7202	7210	7218	7226	7235	52
53	7243	7251	7259	7267	7275	7284	7292	7300	7308	7316	53
54	7324	7332	7340	7348	7356	7364	7372	7380	7388	7396	54
55	7404	7412	7419	7427	7435	7443	7451	7459	7466	7474	55
56	7482	7490	7497	7505	7513	7520	7528	7536	7543	7551	56
57	7559	7566	7574	7582	7589	7597	7604	7612	7619	7627	57
58	7634	7642	7649	7657	7664	7672	7679	7686	7694	7701	58
59	7709	7716	7723	7731	7738	7745	7752	7760	7767	7774	59
60	7782	7789	7796	7803	7810	7818	7825	7832	7839	7846	60
61	7853	7860	7868	7875	7882	7889	7896	7903	7910	7917	61
62	7924	7931	7938	7945	7952	7959	7966	7973	7980	7987	62
63	7993	8000	8007	8014	8021	8028	8035	8041	8048	8055	63
64	8062	8069	8075	8082	8089	8096	8102	8109	8116	8122	64
65	8129	8136	8142	8149	8156	8162	8169	8176	8182	8189	65
66	8195	8202	8209	8215	8222	8228	8235	8241	8248	8254	66
67	8261	8267	8274	8280	8287	8293	8299	8306	8312	8319	67
68	8325	8331	8338	8344	8351	8357	8363	8370	8376	8382	68
69	8388	8395	8401	8407	8414	8420	8426	8432	8439	8445	69
70	8451	8457	8463	8470	8476	8482	8488	8494	8500	8506	70
71	8513	8519	8525	8531	8537	8543	8549	8555	8561	8567	71
72	8573	8579	8585	8591	8597	8603	8609	8615	8621	8627	72
73	8633	8639	8645	8651	8657	8663	8669	8675	8681	8686	73
74	8692	8698	8704	8710	8716	8722	8727	8733	8739	8745	74
75	8751	8756	8762	8768	8774	8779	8785	8791	8797	8802	75
76	8808	8814	8820	8825	8831	8837	8842	8848	8854	8859	76
77	8865	8871	8876	8882	8887	8893	8899	8904	8910	8915	77
78	8921	8927	8932	8938	8943	8949	8954	8960	8965	8971	78
79	8976	8982	8987	8993	8998	9004	9009	9015	9020	9025	79
80	9031	9036	9042	9047	9053	9058	9063	9069	9074	9079	80
N	0	1	2	3	4	5	6	7	8	9	N

Common Logarithms (Continued)

N	0	1	2	3	4	5	6	7	8	9	N
81	9085	9090	9096	9101	9106	9112	9117	9122	9128	9133	81
82	9138	9143	9149	9154	9159	9165	9170	9175	9180	9186	82
83	9191	9196	9201	9206	9212	9217	9222	9227	9232	9238	83
84	9243	9248	9253	9258	9263	9269	9274	9279	9284	9289	84
85	9294	9299	9304	9309	9315	9320	9325	9330	9335	9340	85
86	9345	9350	9355	9360	9365	9370	9375	9380	9385	9390	86
87	9395	9400	9405	9410	9415	9420	9425	9430	9435	9440	87
88	9445	9450	9455	9460	9465	9469	9474	9479	9484	9489	88
89	9494	9499	9504	9509	9513	9518	9523	9528	9533	9538	89
90	9542	9547	9552	9557	9562	9566	9571	9576	9581	9586	90
91	9590	9595	9600	9605	9609	9614	9619	9624	9628	9633	91
92	9638	9643	9647	9652	9657	9661	9666	9671	9675	9680	92
93	9685	9689	9694	9699	9703	9708	9713	9717	9722	9727	93
94	9731	9736	9741	9745	9750	9754	9759	9763	9768	9773	94
95	9777	9782	9786	9791	9795	9800	9805	9809	9814	9818	95
96	9823	9827	9832	9836	9841	9845	9850	9854	9859	9863	96
97	9868	9872	9877	9881	9886	9890	9894	9899	9903	9908	97
98	9912	9917	9921	9926	9930	9934	9939	9943	9948	9952	98
99	9956	9961	9965	9969	9974	9978	9983	9987	9991	9996	99
N	0	1	2	3	4	5	6	7	8	9	N

Table of Sines, Cosines, and Tangents

Angle	Radians	Sine	Cosine	Tangent	Angle	Radians	Sine	Cosine	Tangent
0°	.0000	.0000	1.0000	.0000					
1	.0175	.0175	.9998	.0175	21°	.3665	.3584	.9336	.3839
2	.0349	.0349	.9994	.0349	22	.3840	.3746	.9272	.4040
3	.0524	.0523	.9986	.0524	23	.4014	.3907	.9205	.4245
4	.0698	.0698	.9976	.0699	24	.4189	.4067	.9135	.4452
5	.0873	.0872	.9962	.0875	25	.4363	.4226	.9063	.4663
6	.1047	.1045	.9945	.1051	26	.4538	.4384	.8988	.4877
7	.1222	.1219	.9925	.1228	27	.4712	.4540	.8910	.5095
8	.1396	.1392	.9903	.1405	28	.4887	.4695	.8829	.5317
9	.1571	.1564	.9877	.1584	29	.5061	.4848	.8746	.5543
10	.1745	.1736	.9848	.1763	30	.5236	.5000	.8660	.5774
11	.1920	.1908	.9816	.1944	31	.5411	.5150	.8572	.6009
12	.2094	.2079	.9781	.2126	32	.5585	.5299	.8480	.6249
13	.2269	.2250	.9744	.2309	33	.5760	.5446	.8387	.6494
14	.2443	.2419	.9703	.2493	34	.5934	.5592	.8290	.6745
15	.2618	.2588	.9659	.2679	35	.6109	.5736	.8192	.7002
16	.2793	.2756	.9613	.2867	36	.6283	.5878	.8090	.7265
17	.2967	.2924	.9563	.3057	37	.6458	.6018	.7986	.7536
18	.3142	.3090	.9511	.3249	38	.6632	.6157	.7880	.7813
19	.3316	.3256	.9455	.3443	39	.6807	.6293	.7771	.8098
20	.3491	.3420	.9397	.3640	40	.6981	.6428	.7660	.8391

Table of Sines, Cosines, and Tangents (Continued)

Angle	Radians	Sine	Cosine	Tangent	Angle	Radians	Sine	Cosine	Tangent
41	.7156	.6561	.7547	.8693	66	1.1519	.9135	.4067	2.2460
42	.7330	.6691	.7431	.9004	67	1.1694	.9205	.3907	2.3559
43	.7505	.6820	.7314	.9325	68	1.1868	.9272	.3746	2.4751
44	.7679	.6947	.7193	.9657	69	1.2043	.9336	.3584	2.6051
45	.7854	.7071	.7071	1.0000	70	1.2217	.9397	.3420	2.7475
46	.8029	.7193	.6947	1.0355	71	1.2392	.9455	.3256	2.9042
47	.8203	.7314	.6820	1.0724	72	1.2566	.9511	.3090	3.0777
48	.8378	.7431	.6691	1.1106	73	1.2741	.9563	.2924	3.2709
49	.8552	.7547	.6561	1.1504	74	1.2915	.9613	.2756	3.4874
50	.8727	.7660	.6428	1.1918	75	1.3090	.9659	.2588	3.7321
51	.8901	.7771	.6293	1.2349	76	1.3265	.9703	.2419	4.0108
52	.9076	.7880	.6157	1.2799	77	1.3439	.9744	.2250	4.3315
53	.9250	.7986	.6018	1.3270	78	1.3614	.9781	.2079	4.7046
54	.9425	.8090	.5878	1.3764	79	1.3788	.9816	.1908	5.1446
55	.9599	.8192	.5736	1.4281	80	1.3963	.9848	.1736	5.6713
56	.9774	.8290	.5592	1.4826	81	1.4137	.9877	.1564	6.3138
57	.9948	.8387	.5446	1.5399	82	1.4312	.9903	.1392	7.1154
58	1.0123	.8480	.5299	1.6003	83	1.4486	.9925	.1219	8.1443
59	1.0297	.8572	.5150	1.6643	84	1.4661	.9945	.1045	9.5144
60	1.0472	.8660	.5000	1.7321	85	1.4835	.9962	.0872	11.43
61	1.0647	.8746	.4848	1.8040	86	1.5010	.9976	.0698	14.30
62	1.0821	.8829	.4695	1.8807	87	1.5184	.9986	.0523	19.08
63	1.0996	.8910	.4540	1.9626	88	1.5359	.9994	.0349	28.64
64	1.1170	.8988	.4384	2.0503	89	1.5533	.9998	.0175	57.29
65	1.1345	.9063	.4226	2.1445					

Number	Number^2	$\sqrt{Number}$	$\sqrt{10 \times Number}$	Number^3
1	1	1.000000	3.162278	1
2	4	1.414214	4.472136	8
3	9	1.732051	5.477226	27
4	16	2.000000	6.324555	64
5	25	2.236068	7.071068	125
6	36	2.449490	7.745967	216
7	49	2.645751	8.366600	343
8	64	2.828427	8.944272	512
9	81	3.000000	9.486833	729
10	100	3.162278	10.00000	1,000
11	121	3.316625	10.48809	1,331
12	144	3.464102	10.95445	1,728
13	169	3.605551	11.40175	2,197
14	196	3.741657	11.83216	2,744
15	225	3.872983	12.24745	3,375
16	256	4.000000	12.64911	4,096
17	289	4.123106	13.03840	4,913
18	324	4.242641	13.41641	5,832
19	361	4.358899	13.78405	6,859
20	400	4.472136	14.14214	8,000
21	441	4.582576	14.49138	9,261
22	484	4.690416	14.83240	10,648
23	529	4.795832	15.16575	12,167
24	576	4.898979	15.49193	13,824
25	625	5.000000	15.81139	15,625
26	676	5.099020	16.12452	17,576
27	729	5.196152	16.43168	19,683
28	784	5.291503	16.73320	21,952
29	841	5.385165	17.02939	24,389
30	900	5.477226	17.32051	27,000
31	961	5.567764	17.60682	29,791
32	1,024	5.656854	17.88854	32,768
33	1,089	5.744563	18.16590	35,937
34	1,156	5.830952	18.43909	39,304
35	1,225	5.916080	18.70829	42,875
36	1,296	6.000000	18.97367	46,656
37	1,369	6.082763	19.23538	50,653
38	1,444	6.164414	19.49359	54,872
39	1,521	6.244998	19.74842	59,319
40	1,600	6.324555	20.00000	64,000
41	1,681	6.403124	20.24846	68,921
42	1,764	6.480741	20.49390	74,088
43	1,849	6.557439	20.73644	79,507
44	1,936	6.633250	20.97618	85,184

From Buchsbaum's Complete Handbook of Practical Electronic Reference Data, W. H. Buchsbaum, © 1978, 1973, Prentice-Hall, page 588 to 592.

Table III.1 Number Functions.

Number	Number²	√Number	√10 · Number	Number³
45	2,025	6.708204	21.21320	91,125
46	2,116	6.782330	21.44761	97,336
47	2,209	6.855655	21.67948	103,823
48	2,304	6.928203	21.90890	110,592
49	2,401	7.000000	22.13594	117,649
50	2,500	7.071680	22.36068	125,000
51	2,601	7.141428	22.58318	132,651
52	2,704	7.211103	22.80351	140,608
53	2,809	7.280110	23.02173	148,877
54	2,916	7.348469	23.23790	157,464
55	3,025	7.416198	23.45208	166,375
56	3,136	7.483315	23.66432	175,616
57	3,249	7.549834	23.87467	185,193
58	3,364	7.615773	24.06319	194,112
59	3,481	7.681146	24.28992	205,379
60	3,600	7.745967	24.49490	216,000
61	3,721	7.810250	24.69818	226,981
62	3,844	7.874008	24.89980	238,047
63	3,969	7.937254	25.09980	250,047
64	4,096	8.000000	25.29822	262,144
65	4,225	8.062258	25.49510	274,625
66	4,356	8.124038	25.69047	287,496
67	4,489	8.185353	25.88436	300,763
68	4,624	8.246211	26.07681	314,432
69	4,761	8.306624	26.26785	328,509
70	4,900	8.366600	26.45751	343,000
71	5,041	8.426150	26.64583	357,911
72	5,184	8.485281	26.83282	373,248
73	5,329	8.544004	27.01851	389,017
74	5,476	8.602325	27.20294	405,224
75	5,625	8.660254	27.38613	421,875
76	5,776	8.717798	27.56810	438,976
77	5,929	8.774964	27.74887	456,533
78	6,084	8.831761	27.92848	474,552
79	6,241	8.888194	28.10694	493,039
80	6,400	8.944272	28.28427	512,000
81	6,561	9.000000	28.46050	531,441
82	6,724	9.055385	28.63564	551,368
83	6,889	9.110434	28.80972	571,787
84	7,056	9.165151	28.98275	592,704
85	7,225	9.219544	29.15476	614,125
86	7,396	9.273618	29.32576	636,056
87	7,569	9.327379	29.49576	658,503
88	7,744	9.380832	29.66479	681,472
89	7,921	9.433981	29.83287	704,969
90	8,100	9.486833	30.00000	729,000

Table III.1 (*Continued*)   Number Functions.

Number	Number^2	$\sqrt{Number}$	$\sqrt{10 \times Number}$	Number^3
91	8,281	9.539392	30.16621	753,571
92	8,464	9.591663	30.33150	778,688
93	8,649	9.643651	30.49590	804,357
94	8,836	9.695360	30.65942	830,584
95	9,025	9.746794	30.82207	857,375
96	9,216	9.797959	30.98387	884,736
97	9,409	9.848858	31.14482	912,673
98	9,604	9.899495	31.30495	941,192
99	9,801	9.949874	31.46427	970,299
100	10,000	10.00000	31.62278	1,000,000

Number	$\sqrt[3]{Number}$	$\sqrt[3]{10 \times Number}$	$\sqrt[3]{100 \times Number}$
1	1.000000	2.154435	4.641589
2	1.259921	2.714418	5.848035
3	1.442250	3.107233	6.694330
4	1.587401	3.419952	7.368063
5	1.709976	3.684031	7.937005
6	1.817121	3.914868	8.434327
7	1.912931	4.121285	8.879040
8	2.000000	4.308869	9.283178
9	2.080084	4.481405	9.654894
10	2.154435	4.641589	10.00000
11	2.223980	4.791420	10.32280
12	2.289428	4.932424	10.62659
13	2.351335	5.065797	10.91393
14	2.410142	5.192494	11.18689
15	2.466212	5.313293	11.44714
16	2.519842	5.428835	11.69607
17	2.571282	5.539658	11.93483
18	2.620741	5.646216	12.16440
19	2.668402	5.748897	12.38562
20	2.714418	5.848035	12.59921
21	2.758924	5.943922	12.80579
22	2.802039	6.036811	13.00591
23	2.843867	6.126926	15.20006
24	2.884499	6.214465	13.38866
25	2.924018	6.299605	13.57209
26	2.962496	6.382504	13.75069
27	3.000000	6.463304	13.92477
28	3.036589	6.542133	14.09460
29	3.072317	6.619106	14.26043
30	3.107233	6.694330	14.42250
31	3.141381	6.767899	14.58100
32	3.174802	6.839904	14.73613
33	3.207534	6.910423	14.88806
34	3.239612	6.979532	15.03695
35	3.271066	7.047299	15.18294

Table III.1 (*Continued*)   Number Functions.

Number	$\sqrt[3]{Number}$	$\sqrt[3]{10 \times Number}$	$\sqrt[3]{100 \times Number}$
36	3.301927	7.113787	15.32619
37	3.332222	7.179054	15.46680
38	3.361975	7.243156	15.60491
39	3.391211	7.306144	15.74061
40	3.419952	7.368063	15.87401
41	3.448217	7.428959	16.00521
42	3.476027	7.488872	16.13429
43	3.503398	7.547842	16.26133
44	3.530348	7.605905	16.38643
45	3.556893	7.663094	16.50964
46	3.583048	7.719443	16.63103
47	3.608826	7.774980	16.75069
48	3.634241	7.829735	16.86865
49	3.659306	7.883735	16.98499
50	3.684031	7.937005	17.09976
51	3.708430	7.989570	17.21301
52	3.732511	8.041452	17.32478
53	3.756286	8.092672	17.43513
54	3.779763	8.143253	17.54411
55	3.802952	8.193213	17.65174
56	3.825862	8.242571	17.75808
57	3.848501	8.291344	17.86316
58	3.870877	8.339551	17.96702
59	3.892996	8.387207	18.06969
60	3.914868	8.434327	18.17121
61	3.936497	8.480926	18.27160
62	3.957892	8.527019	18.37091
63	3.979057	8.572619	18.46915
64	4.000000	8.617739	18.56636
65	4.020726	8.662391	18.66256
66	4.041240	8.706588	18.75777
67	4.061548	8.750340	18.85204
68	4.081655	8.793659	18.94536
69	4.101566	8.836556	19.03778
70	4.121285	8.879040	19.12931
71	4.140818	8.921121	19.21997
72	4.160168	8.962809	19.30979
73	4.179339	9.004113	19.39877
74	4.198336	9.045042	19.48695
75	4.217163	9.085603	19.57434
76	4.235824	9.125805	19.66095
77	4.254321	9.165656	19.74681
78	4.272659	9.205164	19.83192
79	4.290840	9.244335	19.91632
80	4.308869	9.283178	20.00000

Table III.1 (*Continued*) Number Functions.

Number	$\sqrt[3]{Number}$	$\sqrt[3]{10 \times Number}$	$\sqrt[3]{100 \times Number}$
81	4.326749	9.321698	20.08299
82	4.344481	9.359902	20.16530
83	4.362071	9.397796	20.24694
84	4.379519	9.435388	20.32793
85	4.396830	9.472682	20.40828
86	4.414005	9.509685	20.48800
87	4.431048	9.546403	20.56710
88	4.447960	9.582840	20.64560
89	4.464745	9.619002	20.72351
90	4.481405	9.654894	20.80084
91	4.497941	9.690521	20.87759
92	4.514357	9.725888	20.95379
93	4.530655	9.761000	21.02944
94	4.546836	9.795861	21.10454
95	4.562903	9.830476	21.17912
96	4.578857	9.864848	21.25317
97	4.594701	9.898983	21.32671
98	4.610436	9.932884	21.39975
99	4.626065	9.966555	21.47229
100	4.641589	10.00000	21.54435

Table III.1 (*Continued*)   Number Functions.

Standard Semiconductor Symbols

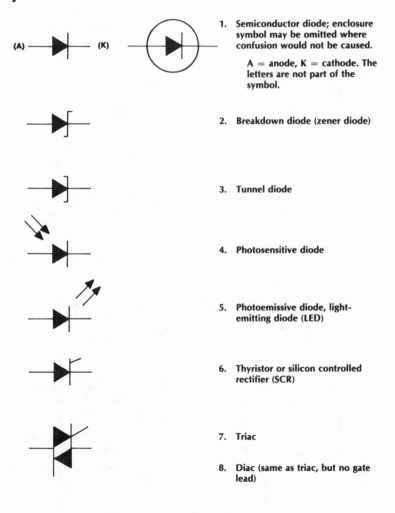

1. Semiconductor diode; enclosure symbol may be omitted where confusion would not be caused.

   A = anode, K = cathode. The letters are not part of the symbol.

2. Breakdown diode (zener diode)

3. Tunnel diode

4. Photosensitive diode

5. Photoemissive diode, light-emitting diode (LED)

6. Thyristor or silicon controlled rectifier (SCR)

7. Triac

8. Diac (same as triac, but no gate lead)

Figure IV.1 Graphic Symbols.

271

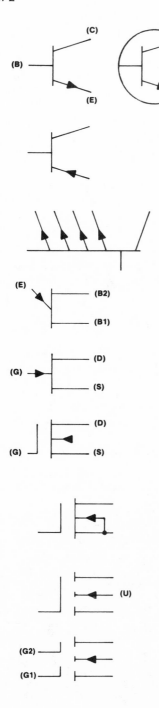

1. **NPN transistor; enclosure symbol may be omitted where confusion would not be caused (unless an electrode is connected to it, as shown here).**

   C = collector, E = emitter, B = base; these letters are not part of the symbol.

2. **PNP transistor**

3. **NPN transistor with multiple emitters (four shown in this example)**

4. **Unijunction transistor with N-type base. If arrow on emitter points in opposite direction base is P type.**

5. **Junction field-effect transistor (JFET) with N-channel junction gate.**

   G = gate, D = drain, S = source; these letters are not part of the symbol.

6. **Insulated-gate field-effect transistor (IGFET) with N-channel (depletion type), single gate, positive substrate.**

7. **Insulated-gate field-effect transistor (IGFET), with N-channel (depletion type), single gate, active substrate internally terminated to source.**

8. **Insulated-gate field-effect transistor (IGFET) with N-channel (enhancement type), single gate, active substrate externally terminated.**

   U = substrate; this letter is not part of symbol.

9. **Same as previous example, but with two gates.**

Figure IV.1 (*Continued*)   Graphic Symbols.

Binary Number System

The binary number system is a positional numeral system employing only two different symbols, 0 and 1. These are called **binary digits**, or **bits**. The importance of the binary number system to information theory and computer technology arises from its ability to represent two-state or bistable conditions such as "on-off," "open-closed," "go-no go," "true-false," and other logic states, which can be simulated by the presence or absence of corresponding electrical potentials.

A positional numeral system has a **base** equal to the number of symbols. In the familiar decimal system, there are ten symbols, 0 through 9. The base is therefore 10. The position a symbol occupies in the expression of a number determines its value, which is the number multiplied by the power of ten corresponding to that position. In the decimal system, the first five positions to the left of the decimal point are units, tens, hundreds, thousands, and tens of thousands; or 10^0, 10^1, 10^2, 10^3, and 10^4. We write the numbers in the opposite direction, of course, but you can easily see that a number such as 12,345 is a shorthand way of writing:

$$1 \times 10^4 + 2 \times 10^3 + 3 \times 10^2 + 4 \times 10^1 + 5 \times 10^0$$

In the binary number system, we replace the base 10 by the base 2 to get position values as follows:

$$\times 2^7 \times 2^6 \times 2^5 \times 2^4 \times 2^3 \times 2^2 \times 2^1 \times 2^0$$

so that a number in binary digits occupying these positions:

$$1\ 0\ 1\ 1\ 0\ 0\ 1\ 1$$

would have the value (in decimal) of:

$$
\begin{array}{rcr}
1 \times 2^7 & = & 128 \\
0 \times 2^6 & = & 0 \\
1 \times 2^5 & = & 32 \\
1 \times 2^4 & = & 16 \\
0 \times 2^3 & = & 0 \\
0 \times 2^2 & = & 0 \\
1 \times 2^1 & = & 2 \\
1 \times 2^0 & = & \underline{1} \\
& & 179
\end{array}
$$

In a binary number, the digit on the right is called the **least significant bit (LSB)**, and that on the left, the **most significant bit (MSB)**. Eight bits are a **byte**, and the four to the right are called **low order bits**, while the four to the left are called **high order bits**.

Addition, subtraction, multiplication, and division are all performed in a manner similar to decimal arithmetic, as in the following examples.

Example 1

Add 1011 and 1100

1011	1st column:	$0 + 1 = 1.$	Write 1.
1100	2nd column:	$0 + 1 = 1.$	Write 1.
———	3rd column:	$1 + 0 = 1.$	Write 1.
10111	4th column:	$1 + 1 = 10.$	Write 10.

Example 2

Subtract 1010 from 10011

10011	1st column:	$0 + ? = 1, 0 + 1 = 1.$	Write 1.
1010	2nd column:	$1 + ? = 1, 1 + 0 = 1.$	Write 0.
———	3rd column:	$0 + ? = 0, 0 + 0 = 0.$	Write 0.
1001	4th column:	Borrow, then $1 + ? = 10,$	
		$1 + 1 = 10.$	Write 1.

Example 3

Multiply 1011 by 101

1011	This is easy. Since we never multiply by any
101	number except 1, it's only a matter of copying
───────	and adding zeros where necessary.
1011	
10110	
───────	
110111	

Example 4

Divide 1000010101 by 1101

       101001

1101 ÷ 1000010101	Either the divisor "goes," or it does
1101	not, so you have only two choices at
────	each step.
1110	
1101	
────	
1101	
1101	
────	

You might like to do these computations in decimal form to check the answers.

The logical circuitry required to perform these operations is quite simple. Figure V.1 shows a **half adder**, consisting of an AND gate and an exclusive-OR gate. Bits to be added are applied at inputs A and B. The truth table summarizes the results at the **sum** and **carry** outputs of various inputs.

The reason for calling this circuit a half adder is that it requires two to handle the carry. Two half adders are one full adder, of course, and Figure V.2 shows a set of four full adders operating in parallel to add the number 0101 and 0011.

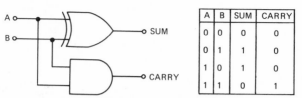

From Encyclopedia of Integrated Circuits, *W.H. Buchsbaum, 1981, Prentice-Hall, page 114.*

Figure V.1   Half Adder.

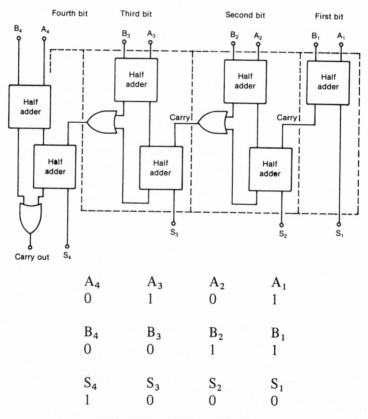

A_4	A_3	A_2	A_1
0	1	0	1

B_4	B_3	B_2	B_1
0	0	1	1

S_4	S_3	S_2	S_1
1	0	0	0

Figure V.2   A 4-bit parallel adder.

Circuits like this one can do addition, subtraction, multiplication and division. To do subtraction, the number that is to be taken away is first passed through an inverter, so that all its 1s are changed to 0s, and all its 0s to 1s. It is then *added*. The carry is carried around and added to the LSB. The answer is the same as in regular subtraction. Try it and see. This method is called the **1s complement**. Another method of subtracting by adding is called the **2s complement**. In this the carry is ignored, but an extra 1 is added to the LSB anyway. It's really a slightly different circuit rather than a different method.

Multiplication is performed by repeated addition, and division by repeated subtraction. This might seem tedious to us, but to a computer operating at megahertz speeds, it is a negligible consideration.

Index

A

α (see alpha)
Absolute zero, 22
Acceptor impurities, 24
  aluminum, 24
  boron, 24
  indium, 24
Accumulator, 116
Active filter, 154
ADC (see analog-to-digital converter)
Adder, 275
  full adder, 275
  half adder, 275
Address bus, 116
Alpha, 37
Alpha cutoff frequency, 40
ALU (see arithmetic logic unit)
Aluminum, 24
Analog-to-digital converter, 149
AND gate, 57
Anode, 35
Antimony, 23
Arithmetic logic unit, 116
Arsenic, 23

B

β (see beta)
Base, 36
BCD counter (see binary-coded decimal counter)
Beryllium oxide, 161
Beta, 39
Beta cutoff frequency, 40
Bias, 35
Bilateral symmetry, 28
Binary-coded decimal counter, 94
Binary digits, 57, 273
Bipolar device, 24
Bipolar fabrication, 47

Bipolar junction transistor, 34
  alpha, 37
  base, 36
  basic structure, 40
  beta, 39
  beta cutoff frequency, 40
  characteristic curves, 38
  collector, 36
  common-emitter circuit, 37
  emitter, 36
  fabrication, 47
  frequency response, 39
  junctions, 35
  load line, 39
  npn, 36
  operating point, 39
  pnp, 36
  used as diode, 40
  used as resistor, 40
Bistable mulivibrator
  (see also flip-flop), 85
Bits (binary digits), 57, 273
BJT (see bipolar junction transistor)
Bonding, 50
  thermocompression ("nailhead"), 50
  ultrasonic, 53
Boron, 24
Bubble memory, 110
Buffer, 96
Buried layer, 47
Byte, 120, 236

C

Capacitors in ICs, 41
Carrier mobility, 29
  carrier mobility unit, 30
  drift velocity, 30
  electron mobility, 32
  hole mobility, 32
  Miller index, 31

Cathode, 35
Cathode-ray tube, 191
Central processing unit, 114
CCD (see charge-coupled device)
Charge-coupled device, 109
Chip, 21
Circulator, 169
Clock, 85, 90
CML (see current-mode logic)
CMOS (see complementary MOS)
CMOS inverter, 60
Collector, 36
Common-emitter circuit, 37
Common-mode rejection, 133
Comparator, differential, 140
Complementary MOS, 79
Computer, 113
   central processing unit, 114
   disk drive control IC, 119
   hardware, 114
   I/O devices, 119
   keyboard interface IC, 122
   memory, bulk, 119
   memory, internal, 105
   microprocessor, 114
   printer interface IC, 125
   software, 114
   video display interface IC, 124
Conduction band, 22
Consumer radar, 172
Control bus, 116
Control circuits, industrial, 141
Counter, 92
   BCD counter, 94
Covalent bonding, 21
CPU (see central processing unit)
CRT (see cathode-ray tube)
Crystal lattice, 22
Current-mode logic, 73
Current-voltage characteristics of MOSFETs,
   32
Czochralski method of crystal pulling, 42

D

DAC (see digital-to-analog converter)
Darlington amplifier, 131
Data buffer, 116
Data bus, 116
DC drift, 131

DCTL (see direct-coupled transistor logic)
Decoder, 99
Delay lines, 170
Delta modulation, 179
Demodulation, optical, 183
Depletion layer, 27
Depletion-mode MOSFET, 27
Detectors, 145
D flip-flop, 87
Differential amplifier, 132
Diffusion, 46
Digital circuits, 57
   accumulator, 116
   BCD counter, 94
   bistable multivibrator, 85
   bubble memory, 110
   buffer, 96
   decade counting unit (see BCD counter)
   decoder, 99
   display driver, 99
   encoder, 97
   flip-flop, 85
   full adder, 275
   half adder, 275
   inverter, 58
   shift register, 91
   transistor switch, 57
Digital-to-analog converter, 152
Diode, 35
Diode-array IC, 155
Diodes in ICs, 40
Diode-transistor logic, 70
DIP (see dual-in-line package)
Direct-coupled transistor logic, 68
Direct coupling, 130
Disk drive controller IC, 119
Display driver, 99
Displaying information, 186
   CRT, 191
   LCD, 188
   LED, 188
   gas discharge, 191
   vacuum-fluorescent, 190
DM (see delta modulation)
Donor impurities, 24
   antimony, 23
   arsenic, 23
   phosphorus, 23
Dopant, 30
Drain, 25
Drift velocity, 30

DTL (see diode-transistor logic)
Dual-in-line package, 50

E

EAROM (see electrically alterable read-only memory)
ECL (see emitter-coupled logic)
Electrically alterable read-only memory, 109
Electron-beam machining, 49
Electrons, 21
  at junction, 35
  charge carriers, 24
  conduction, 22
  electronvolt, 239
  free, 23
  holes, 23
  in valence shells, 22
  mobility, 29
  thermal energy, 22
  valence, 21
Electronvolt, 239
Electrostatic field, 22
Emitter, 36
Emitter-coupled logic, 73
Energy gap, 22
Encoder, 97
Enhancement-mode MOSFET, 25, 26
Epitaxy, 47
EV (see electronvolt)
Exclusive-OR or NOR gate, 65
Extrinsic semiconductor, 22

F

f_α (see alpha cutoff frequency)
f_β (see beta cutoff frequency)
Fabrication of ICs, 42
  bonding, 50
    thermocompression, 50
    ultrasonic, 53
  Czochralski process, 42
  diffusion, 46
  electron-beam machining, 49
  etching, 46
  floating zone process, 43
  ion implantation, 49
  mounting, 50
  packaging, 50

Fabrication of ICs (cont.)
  performance testing, 50
  photomask, 45
  photoresist, 45
  separation, 50
  SiO$_2$ layer, 44
  vapor epitaxy, 47
  wafer, 44
Fan-in, 67
Fan-out, 67
Feedback volume control, 154
FET (see field-effect transistor)
Fiber optics, 176
  acceptance angle, 179
  cladding, 177
  delta modulation, 179
  demodulation, 184
  flexiscope, 180
  graded-refractive-index fiber, 177
  Integrated Services Digital Network
      (ISDN), 175, 185
  light detector, 182
  light source, 180
    laser, 180
    light-emitting diode, 181
  modulation, 183
    delta modulation, 179
    pulse-coded modulation, 177
  multimode fiber, 177
  normalized frequency, 177
  numerical aperture, 180
  optical communication, 175
  optical spectrum, 176
  pulse-coded modulation, 177
  single-mode fiber, 177
  thermoelectric cooling, 184
  wavelength division multiplexer (WDM),
      185
Field-effect transistor, 24
Field-threshold voltage, 76
Flatpack, 50
Flip-flop, 85
  BCD counter, 94
  bistable multivibrator, 85
  buffer, 96
  counter, 92
  D flip-flop, 87
  J-K flip-flop, 89
  J-K master-slave flip-flop, 89
  R-S flip-flop, 85
  shift register, 91

Floating-zone method of crystal preparation, 43
Forbidden band, 22
Forward bias, 35
Free carrier density, 27
Frequency response
  BJT, 39
  MOSFET, 34

G

Gallium arsenide, 22
Gas discharge display, 191
Gate, 25, 57
Gate, logic,
  AND, 57
  DCTL, 68
  DTL, 70
  ECL, 73
  I^2L, 74
  NAND, 63
  NOR, 63
  NOT, 57
  OR, 57
  RTL, 70
  TTL, 70
  X-NOR, 65
  X-OR, 65
Gate potential, 27
Germanium, 21
Graded-refractive-index fiber, 177
Gunn-effect diode,164

H

Half-adder, 275
Hole-electron pair, 26
Holes, 23
Hydrofluoric acid, 46

I

IC (see integrated circuit)
IF amplifier, 139
  narrowband, 140
  wideband video, 139
IGFET (see insulated-gate field-effect transistor)

IIL or L^2L (see integrated injection logic)
Impact avalanche transit time diode, 163
IMPATT (see impact avalanche transit time
  diode)
Impurities, 22, 23
Indium, 24
Inductors in ICs, 42
Injection laser, 180
Input/output (I/O) device, 114, 119
Insulated-gate field-effect transistor, 25
Integrated circuit, 21
Integrated injection logic, 74
Integrated Services Digital Network, 175,
  185
Intrinsic semiconductor, 22
Inversion layer, 27
Inverter, 58
Ion implantation, 49
ISDN (see Integrated Services Digital Net-
  work)

J–K

JFET (see junction field-effect transistor)
J-K flip-flop, 89
  master-slave, 89
Junction field-effect transistor, 25, 28
Junctions, 35
Keyboard interface IC, 122

L

Latch, 85
LCD (see liquid crystal display)
LED (see light-emitting diode)
Level shifter, 100
Light-emitting diode, 181, 188
Limited space-charge accumulation diode,
  166
Linear circuits, 57, 129
  active filter, 154
  analog computer, 134
  analog-to-digital converter, 138, 149
  audio detector, 145
  common-mode rejection, 133
  Darlington amplifier, 131
  direct-coupled amplifier, 129
    Darlington, 131

Linear circuits (*cont.*)
  direct-coupled amplifier (*cont.*)
    npn amplifier, 131
    npn/pnp amplifier, 131
    DC drift, 131
    differential amplifier, 132
    differential comparator, 140
    digital-to-analog converter, 152
    diode array IC, 155
    feedback volume control, 154
    industrial controls, 141
    narrowband IF amplifier, 140
    operational amplifier, 134
      analog-to-digital converter, 138
      averaging amplifier, 138
      closed-loop gain, 136
      external negative feedback, 135
      integrator, 138
      inverting input, 136
      monostable multivibrator, 138
      noninverting input, 136
      open-loop gain, 136
      rectifier, 138
      sweep generator, 138
    phase-locked loop, 147
    power amplifier, 143
    video detector, 145
    voltage-controlled oscillator (VCO), 147
    voltage regulator, 148
    wideband video amplifier, 139
Line driver, 96
Liquid crystal display, 188
Load line, 39
Logic circuits, 57
  AND gate, 57
  BCD counter, 94
  bistable multivibrator, 85
  CMOS inverter, 60
  decade counting unit, 94
  decoder, 99
  display driver, 99
  encoder, 97
  flip-flop, 85
  full adder, 275
  half adder, 275
  inverter, 58
  NAND gate, 63
  NOR gate, 63
  NOT gate, 57
  OR gate, 57

Logic circuits (*cont.*)
  X-NOR gate, 65
  X-OR gate, 65
Logic families, 67
  DCTL, 68
  DTL, 70
  CML, 73
  CMOS, 79
  ECL, 73
  IIL or I^2L, 74
  NMOS, 79
  PMOS, 75
  RTL, 70
  SOS, 81
  STL, 75
  TTL, 70
LSA (see limited space-charge accumulation
  diode)

M

Magnetic disk bulk memory, 119
Majority carriers, 24
Memory, 57
  bubble, 110
  charge-coupled device (CCD), 109
  electrically alterable read-only memory
    (EAROM), 109
  flip-flop, 85
  programmable read-only memory (PROM),
    109
  random-access memory (RAM), 105
  refreshing, 107
  read-only memory (ROM), 105
  shift register, 91
Metal-oxide-semiconductor fabrication, 44
  (see also fabrication of ICs)
Metal-oxide-semiconductor field-effect transistor, 25
Microacoustic line, 162
Microalloy, 46
Microelectronic radar system, 169
Microprocessor, 113
  accumulator, 116
  address buffer, 115
  address bus, 116
  arithmetic logic unit (ALU), 116
  central processing unit (CPU), 114

Microprocessor (*cont.*)
  control bus, 116
  data buffer, 116
  data bus, 116
  instruction decoder, 115
  instruction register, 115
  program counter, 115
  status flag, 116
  stack pointer, 117
Microstrip, 161
Microwave devices, 159
  automobile radar, 173
  beryllium oxide, 161
  circulator, 169
  dispersive delay line, 170
  dispersive filter, 170
  Gunn-effect diode, 164
  IMPATT diode, 163
  low-noise amplifier (LNA), 172
  LSA diode, 166
  microacoustic line, 162
  microstrip, 161
  phased-array antenna, 167
  phase shifter, 171
  PIN diode, 171
  pulse-compression filter, 168
  radar systems, 160, 169
  RF down converter, 172
  satellite reception, 172
  Schottky barrier diode, 164
  varactor, 163
Miller index, 31
Millisiemens, 33
Minority carriers, 24
Modulation, optical, 183
MOS fabrication (see metal-oxide-semicon-
  ductor fabrication)
MOSFET, MOS/FET (see metal-oxide-
  semiconductor field-effect transistor)
MOSFET and BJT compared, 82
Multimode fiber, 177
Multiplexer, 100

N

NAND gate, 63
Nitride-metal-oxide-semiconductor (NMOS)
  device, 79
NMOS (see nitride-metal-oxide-semiconductor device)

Noise margin, 66
NOR gate, 63
NOT gate, 57
Npn, 35
N-type semiconductor, 23

O

Ohmic joint, 46
Operating point, 39
Operational amplifier, 134
  analog-to-digital converter, 138
  averaging amplifier, 138
  closed-loop gain, 136
  external negative feedback, 135
  integrator, 138
  inverting input, 136
  monostable multivibrator, 138
  noninverting input, 136
  open-loop gain, 136
  rectifier, 138
  sweep generator, 138
  testing, 232
Optical spectrum, 176
Optoisolators, 193
OR gate, 57

P–Q

Packaging, 50
P-channel, 25-27
P-channel metal-oxide-semiconductor device,
  75
PCM (see pulse-code modulation)
Perfboard, 203
Performance testing, 50
Phased-array antenna, 167
Phase-locked loop, 147
Phase shifter, 171
Phosphorus, 23
Photomask, 45
Photoresist, 45
Pinch-off voltage, 22
PIN diode, 182
PMOS (see p-channel metal-oxide-semicon-
  ductor device)
Pnp, 35
Point-contact transistor, 36
Polycrystalline silicon, 42

Power amplifier, 143
Printed-circuit board, 195
  making your own, 199
  manufacturing, 197
  repairs to, 204
  types, 195
Printer interface IC, 125
Program counter, 116
Programmable read-only memory, 109
PROM (see programmable read-only
  memory)
Propagation delay, 67
P-type semiconductor, 23
Pulse-code modulation, 149, 151
Pulse-compression filter, 168
Pulses, 57

R

Radar, 159
RAM (see random-access memory)
Random-access memory, 105
Read-only memory, 105, 108
Recombination, 35
Rectifier, 148
Refreshing, 107
Resistors in ICs, 40
Resistor-transistor logic, 70
Reverse bias, 35
ROM (see read only memory)
R-S flip-flop, 85
RTL (see resistor-transistor logic)

S

Saturation region, 34
Schottky transistor logic, 75
Semiconductor, 21
Separation of wafer, 50
Shells (in atom), 21
Shift register, 91
Silica, 42
Silicon, 21
Silicon dioxide (SiO_2), 25, 44
Silicon nitride gate process, 77
Silicon-on-sapphire device, 81
Silicon tetrachloride, 42
Silicon wafer, 42
Single-crystal silicon, 42

Single-mode fiber, 177
SiO_2 (see silicon dioxide)
Solid-state display screen, 191
SOS (see silicon-on-sapphire)
Source, 25
Speed/power product, 76
Stack pointer, 117
Status flag, 116
STL (see Schottky transistor logic)
Switching transistors, 57

T

Temperature effects, 24
Test equipment for digital ICs, 210
Thermal energy, 22
Thermocompression ("nailhead") bonding, 50
Threshold voltage, 27, 65
Timer, 101
TO metal can package, 50
Transconductance of MOSFET, 32
Transistor
  bipolar, 34
  unipolar, 24
Transistor-transistor logic, 70
Triode region, 32, 33
Troubleshooting equipment with digital ICs,
  207
  common defects, 208
  isolating defect, 215
  logic clip, 212
  logic comparator, 213
  logic probe, 212
  logic pulser, 212
  logic state analyzer, 218
  microprocessor, 217
Troubleshooting equipment with linear ICs,
  223
  brute force, 231
  component tests, 231
  DC voltage analysis, 227
  part substitution, 231
  resistance checks, 225
  signal injection, 230
  signal tracing, 227
  visual inspection, 225
  waveform analysis, 229
Truth table, 59
TTL (see transistor-transistor logic)

U–Z

Ultrasonic bonding, 50
Unipolar device, 24
Vacuum-fluorescent display, 190
Valence electrons, 21
Vapor epitaxy, 47
Varactor, 163
V_G (gate voltage), 27

Video amplifier, 139
Video display interface IC, 124
Voltage-controlled oscillator (VCO), 147
Voltage regulator, 148
V_p (pinch-off voltage), 28
V_T (threshold voltage), 27
Wafer, 42
Wire wrapping, 204